SIMON

Also by Alexander Masters

Stuart: A Life Backwards

ALEXANDER MASTERS

Simon

The Genius in My Basement

Delacorte Press • *New York*

Published in the United States by Delacorte Press,
an imprint of The Random House Publishing Group,
a division of Random House, Inc., New York.

DELACORTE PRESS is a registered trademark of Random House, Inc.,
and the colophon is a trademark of Random House, Inc.

Originally published in hardcover in the United Kingdom by Fourth Estate,
an imprint of HarperCollins Publishers Limited, as *The Genius in My Basement*,
in 2011.

Library of Congress Cataloging-in-Publication Data

Masters, Alexander.
Simon: the genius in my basement / Alexander Masters.
p. cm.
ISBN 978-0-385-34108-0
eBook ISBN 978-0-345-53221-3
1. Norton, Simon. 2. Mathematicians—England—Biography.
3. Finite simple groups. I. Title.
QA29.N67M37 2012
51092—dc23 2011032690
[B]

Printed in the United States of America on acid-free paper

www.bantamdell.com

2 4 6 8 9 7 5 3 1

First Edition

For gorgeous Flora

Oh dear, I have a feeling this book
is going to be a disaster for me.
Simon Norton

SIMON

1

Simon was one year old, playing in the dining room, getting under his mother's stilettos.

He was unusually thoughtful. His brothers at this age pounded the toy blocks on the glass coffee table and jabbed them into the electric sockets.

Simon picked up a pink block from the pile beside his knee and smoothed it against the carpet. Carefully, he positioned a blue brick alongside. He reached across—his mother, on her way to lay the side plates and forks, had to make a sharp swerve—for two more pink bricks, and slid them against the blue. With precision, he extracted another blue brick.

Shuffling across the room on his bottom, Simon found four more pink bricks, fumbled them back and continued the arrangement.

His mother, halfway through folding napkins into bishops' miters, stopped in astonishment. She saw at last what he was doing.

One blue, one pink.

One blue, two pinks.

One blue, three pinks.

One blue, four pinks.

From the disarray of Nature, her baby son was enforcing regularity.

It took our species from the birth of prehistory to the dawn of Babylonian civilization to learn mathematics.

Simon was bumping about its foothills in just over twelve months.

At three years, eleven months and twenty-six days, he toddled into cake layers of long multiplication:

(January 1956)

Simon's brother Francis had barely managed to recite the digits from one to ten by the time he was four years old; his brother Michael, a fraction quicker, had understood that if you gave him three banana-flavor milkshakes, and asked him to "count" them, the correct answer was "one" for the first, "two" for the second and "three" for the sticky splosh dribbling down his ear.

Percentages, square numbers, factors, long division, his 81 and 91 times tables, making numbers dance about to itchy tunes:

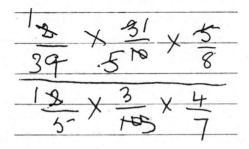

Simon mastered these when he was five.

Occasionally, his attention wandered:

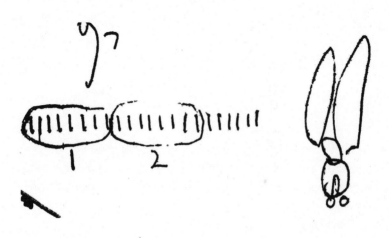

2 The reader meets Simon

Sschliissh, dhuunk, dhuunk, zwaap, dhuunk, zwaap . . .

Listen! Can you hear?
. . . dhuunk, sschliissh, dhuunk, zaap, zwap, dhuunk . . .
Bend down. Put your ear against the carpet:
Zwaap, dhuunk, dhuunk, dhuunk,
zwaap, dhuunk, ssschliissh . . .
It's fifty years later.
. . . liissh,
dhuunk,
dhuunk,
dhuunk,
zwaap,
dhuunk,
ssschliissh,
dhunnk
dhuunk, zwa
ap, sclissh
dhunnk, du
unnk, s

That's the sound of a once-in-a-generation genius.
Simon Phillips Norton: Phillips, with an "s," as if one Phillip were not enough to contain his brilliance. He lives under my floorboards.
Dhuunk, dhuunk . . .

When I first moved here, I had no idea what the noises were. Underground rivers? The next-door neighbors dragging a new pot through to their Tuscan garden? *Dhuunk, dhuunk . . .* But after eight years of interpretation I know that it's the great man's feet, stomping from one end of his room to the other. Every second stomp is heavier.

"*Ssschlissh*": that's the swipe of his puffa ski jacket against the stalagmites of paperbacks he keeps piled on the furniture.

"*Zwaap*": the sound of his duffel, as he rotates at the end of the room. He sometimes flings it wide, hitting papers. Simon carries this bag about with him everywhere he goes, clutched in the crook of his arm, even if it's just to his front door to let in the gas man.

. . . dhuunk, dhuunk, dhuunk, zwaap, dhuunk, dhuunk . . .

Simon's bed is ten feet directly beneath mine. My study is on top of his living room. His stomping space extends the full depth of the building, under my floor. My balcony is the roof of his basement extension, which has herded all the pretty garden plants into a six-foot square at the back of our house and stamped them under concrete slabs.

The phone rings. A charge from Simon: *Dhuunk! Dhuunk! Dhuunk!*

Snorting. The receiver—. . . *rrinng*, clank, clumpump, *ping, ping . . .*—wrenched from its holster: Attempts at speech, grunts, bangs of talk-noise; a strangulated word.

Clunk. Phone back in its holster.

Silence.

Dhuunk, dhuunk, dhuunk . . .

There's another very important sound, which is too difficult to represent typographically: an intermittent, twisted crackle, sharp but thick, with a strong sense of command, resting on a base of plosive disorder. In an exercise book from when he was five there's a squiggle that comes close:

It's the sound of plastic-bag-being-opened-in-a-hurry-and-the-gratification-of-discovering-important-papers-inside. Without understanding this noise, you cannot understand the man.

... *ssschliissh, dhuunk, zwaap, zwaap, dhuunk, dhuunk* ...

Simon has been pacing down there for twenty-seven years, three months, five days, thirteen hours and eight minutes.

Ssssh!

      Stop breathing!

             Did you catch that?

                    Still another sort of noise?

                        A sort of sigh?

                        That was a thought.

Minus N

Your representation of me as interesting is
inaccurate. I feel ashamed by it.

Simon

Damn! He's gone!

Simon's refused to enter the book!
He is a Minus Norton.
"Why now?" I demanded, jumping up from the carpet when
he stomped into my study from the basement. "The reader has
started the story. He's spent the money. He feels conned."
"How do you know it's a he who's reading it? It might be a
she, hnnn."
"He or she! Who cares?"

"I presume they do," he said cunningly.

Behind him, a bubble of air floated up the stairs and expanded into my rooms of the house, whiffing of damp and sardines.

Then he barged out of the front door, and, the scuff of his sandals becoming rapidly soft and seaside-ish, disappeared toward the Mathematics Faculty.

A book about Simon that doesn't have Simon in it?

I had thought a life of Simon would be tiptoeing on the edge of the shadow of God. Instead, he crashes about my study as though heel joints had never been invented; makes women shriek when they turn on the light in the corridor and find him standing there like an Easter Island statue; his duffel twists him into animal shapes; he hides behind envelopes.

He shocks me awake with his snores.

Writing biographies of living people, the subject is an irritant. Why is he needed? All he does is insist that whatever you've written is wrong.

In fact, when Simon *was* part of the book, I had to run away from him.

Wouldn't all biographies be better if they gave up trying to fix the person they're writing about, and confined themselves to his glints and reflections—a biography not of Simon but of the perception of Simon? What is a biography, anyway? A platter of gossip and titbits. It's up to the readers to mix these components together in whatever way they find most entertaining and instructive. The subject's out of it. Once word hits page, he's irrelevant.

I'm glad Simon's gone. Good riddance!

In mathematics, you jump onto the subject of numbers through your experience of reality—two flies multiplied by four sudden pulls gives eight wings; three toads, two frogs and one bathtub equals six screams of fury from your father; four bags of crisps and five of your mum's cigarettes make nine orders of stomachache—that's how the newcomer gets introduced to the subject, via the positive, whole numbers: 1, 2, 3, 4, 5, 6 . . .

But mathematicians insist that negative numbers are equally real. It's just a matter of which way you happen to look: going ahead is positive, and going behind is negative.

I'll go behind Simon. Allow me to introduce Biographical Minus N:

Simon Phillips MINUS Norton.

Now, let's break into his basement.

4

November 26, 1922: Carter pierced a small hole in the
wall through which he could look into the Pharaoh's
chamber with a sliver of torch light. Asked if he could
see anything he replied, "yes, wonderful things!"

Howard Carter's discovery of Tutankhamen's tomb

But I can't find the light switch.

Which is important when you're
standing at the top of Simon's
stairs with nothing but sardine stench
and book writer's bones to break your
fall. Every other house in our building has a light switch by the
stairwell door—why has Simon wrenched his out?

It makes me tense. My nerves clench into a knot. It feels
planned.

There are holes in the stair carpet: lips of fabric at the edge
of the treads, cut to flop forward, snatch . . . tap-tap . . . your
toes . . .

. . . and plunge you onto the quarry tiles at the bottom.

These stairs are booby-trapped—against biographers.

It's safest to take the rest of the steps spread-eagle fashion, one foot slithering against the wall while the other rat-a-tats along the banister spindles. The palms of my hands catch and release on splodges of stickiness. As I slide down, I pass over two treads that have been blasted away. The wood has been broken in. It's a sheer drop between the thigh-shredding splinters left behind to the floor below. Craftily, Simon has left the carpet in place over the chasm.

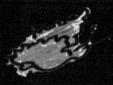

Phhuuuuh! What was that? A moth? No. Just a grease dollop drifting by.
Unidentified species often float up this stairwell.

The only person who has been caught by this booby trap is the booby who manufactured it in the first place, Dr. Simon MINUS Norton. The other week, I remember, I saw him leaping about the street on one leg, clutching his knee.

At last, here, at the bottom of the steps, we encounter a switch . . .

The bulb—low-watt, energy-saving—spreads shadow, not light.

It gathers a narrow entrance lobby into view, the floor of which is strewn with wood shavings and brick fragments. Sections of plaster have chipped away from the walls, exposing shoddy Victorian masonry. Along one edge of one side of the carpet is a pile of merry-colored supermarket bags—perhaps forty in total, traffic-light orange, Pacific blue, lime-green stripes—the plastic straining colonicly against the mass of paperwork rammed inside.

If we squeeze over the rubble and past these plastic bags, we can peer through a door frame that appears to have lost its door. Wrinkle your nose. Squint your eyes. This is Simon's basement: long, low and odoriferous.

There are so many words Simon refuses to let me use:

"S—" (seven letters, including a "q.")

"Too scandalous!"

"P—" (six letters, oink, oink.)

"My poor mother!"

"C—" (seven, mild, rhymes with butter.)

"How shaming!"

"M—" (six, obscure, but not to Simon; investigated by archaeologists.)

"Stop writing immediately!"

Simon's Banned List is a page and a half long. Our most violent argument was over the four-letter "f—" word.

"No!" he strangulated.

I am not to use this "f—" . . .

"No!" he wriggled.

. . . to describe Simon's fraction of the house under any circumstance. This word "f—" . . .

"No!" he sank piteously to his knees.

will get him into trouble with the police.

What am I to say?

"Rooms," was Simon's genteel proffering.

"No!" I started from my writing chair. "Too polite. I'm not going to lie to my readers to *that* extent."

"You've shown no compunction about much greater lies elsewhere."

"But," I relaunched the argument for "f—," "when the house was being assessed for council tax, at one stage the council maintained that it was a separate 'f—.'"

"And it would have meant a lot extra on my council tax bill. Hnnnh, I don't want to have to go through that again, hnnh."

"How about 'apar—'?"

"No! No! No!"

"Bedsit?"

"No!" we shrieked together, and fell about laughing.

Simon has lived in this . . . this . . . this . . . *excavation* since 1981. Once your eyes have adjusted to the gloom, you'll see that it's made up of two rooms: a main one, which extends the full depth of the house, thirty feet from end to end, and the 1970s school-block type of extension at the back that ends with a set of sliding doors opening onto brambles.

Now, slip on your . . . no, wait. I must say something first about the "Titanic Toilet."

Underneath the booby-trapped stairs we just slid down to get here is a corpse—a dead and rotting lavatory bowl.

Simon was sitting on this toilet when the floor gave way. He and the crapper fell into the abyss so fast that his teeth bit his nose and he would have vanished altogether had the underside of the bowl not banged to a stop against the waste pipe and balanced there, beached, holed, the *Titanic* of Toilets, teetering over the center of the earth. Simon hasn't been able to go near the place since, except to "stand." Wedging his head against the low, sloped ceiling of the stairs, clutching the washbasin with both hands, he teeters his toes to the edge of the broken woodwork—and waters the blackness.

When Simon wants "to sit" he considers even my bathroom upstairs too close to the scene of trauma; he has to go to the farthest possible alternative accommodation in the house: the toilet on the top floor.

Returning to the Excavation. Now is the correct moment to slip on your steel boots, belt up with climbing robes and G-clips, grab a few plasters and a bottle of antiseptic: we're about to enter the first cave.

It's easier in here to describe where the paper, plastic bags and books are not than where they are: they're not on the ceiling.

I suppose you could say, technically, there are no papers on the top third of the walls.

A lot of it trembles in towers on the arms of chairs, on tables, on cupboards, on top of a dinner lady's cart that Simon's managed to wrench out of some local school and rattle back along the midnight streets.

There are outlines of walls, outcrops suggesting a clothes cupboard, a padded chair, one, two, possibly three chests of drawers; no discernible floor; and—watch out!—an I-beam thrusting across the ceiling, indicating that, at some point in this cave's history, primitive inhabitants have knocked out a wall, possibly during the Cambridge population explosion of the early 1900s.

Finally, here is floor.

We can rest for a moment now and take our bearings. To the right, the front of the house: a bay window, light cut off by blinds (pattern of blinds: colored, wave-like stripes, perhaps reminding cave inhabitants of the sea). To the left: dinge.

Patterns made in the shadows by the stalagmites of plastic bags and books include a galloping cow, a beetle trying to hide.

under what was once the padded armchair, the face of a grotesque man . . .

Gullies molded into the floor surface of paper-filled supermarket bags, envelopes, squashed boxes, fallen books, mark the route Simon takes when he stomps from his mattress to the toilet under the stairs; the toilet to the kitchen; kitchen (carrying a plate of sardines) back to mattress; then (plate and cutlery bouncing as he leaps up) mattress to front patio door, where the paper disorder gives way and a solitary rut leads through the leaves and fallen shards of masonry big enough to kill, up to the street outside our house—at which point mechanized sweepers and dustbin men take over and the trail disappears.

Amazingly, there are some good pieces of furniture here. Simon's bed (again, to the right, beside the seascape window blinds) is a nineteenth-century mahogany antique with a Dutch gable headboard and cusped legs. How did this splendid object get in? A gift from ancient gods? It's placed so Simon MINUS Norton can hear the postman coming up the steps to the front door of the main floor above, leap out of bed, race up the booby-trap stairs and be trembling there with his hands held in a scoop below the letterbox before the first envelope hits the floor.

An eighteenth-century twizzle-legged side table is at the other end of this first room, by the kitchen, washed up on a patch of carpet: the sort of thing found in Cotswold cottages with an arrangement of desiccated flowers sliding off the edge. This table is at an end point of Simon's stomping route, and bears the brunt of his swinging bag when he turns. Yet somehow a giraffe-shape of paperbacks with alternating dark and light spines has managed to remain in place.

Dr. Simon MINUS Norton likes to read books in a single day—packing in pages on train journeys; during delays at bus stops; between bites of sardines in tomato sauce; while floating in the bath—until he has drained the book of information, after which the broken, dog-eared volume is evicted onto a table, under a cushion, inside a saucepan, and begins a descent, measured in a timescale of years, to the archaeological strata on the floor.

Now that we're inside this first room or cave, we can take off our climbing gear and start closer investigations. At the bottom of the giraffe pile is a ring-bound book, half an inch thick, pillarbox red, the size of a tea tray: *Atlas of Finite Groups*, one of the greatest mathematical publications of the second half of the twentieth century. It's got Simon's name on it.

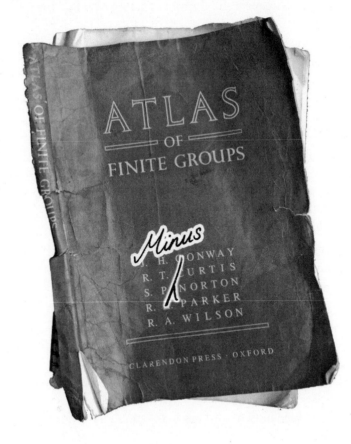

Under the bed, if we push aside *As the Crow Flies,* by Janet Street-Porter, we find a limerick written in a tiny, bumpy hand:

> A young girl of Welwyn, named Helen
> Was playing near a well, and fell in
> She was soaked head to toes
> Was (the question arose)
> Helen well in the well in Welwyn?

It's on school notepaper, cut with strangling precision around the words and folded so that the creases form a tessellation of diamonds.

Venturing your hand in farther under the bed—that's it, up to the elbow—press your chin hard against that mahogany panel . . . there! A slipper. Look inside that: another piece of stationery, this time typed, from the Senior Tutor at Trinity College, Cambridge, dated October 8, 1969:

Dear Norton,
The *Cambridge Evening News* have just been on the line asking whether they could have a few words with you and have a photograph taken. I have told them that you have already been interviewed by the National Press but typically they would want to do it all again . . .

What's this? The slipper is decorated with a small yellow lozenge at the toe end, showing a smiley cartoon face, the symbol for Ecstasy rave parties in the 1990s. Squashed at the end of the slipper is an envelope, crumpled, containing a thick, pressed chunk, a cake of . . .

Splendid! A slab of tooth impressions!

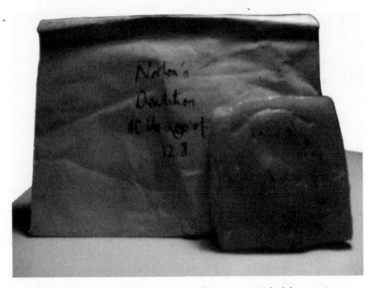

Norton's Dentition At the age of 12.8. Fluorescent pink slab, wax, 3 cm x 3 cm x 0.5 cm. Excavated and photographed by the author, from a slipper.

Flicking at the plastic bags; clucking to ourselves over the titles of books piled on the armchair; scowling at the three disgraceful jackets coated in mold in the clothes cupboard—we clamber about these rooms feeling annoyance. It's hard to put a finger on the reason for it.

It's the vapidity of 99 percent of this junk. If it was only totally vapid, we could dismiss the man's life and move on. But over here, if we climb across two cubic cardboard boxes and slide down the other side of a slope of Asda bags next to this chest of drawers containing Simon's collection of used Tango bottles from the late 1980s, is a second letter. Dated 1971, it's typed with a heavier hand—a fierce attack on the keyboard. In places the letter "o" has come with the center shaded and periods have pierced violently through the paper.

Dear Sir,
As you must be one of the cleverest people alive today, I
wonder if you would be interested in assisting me with a
project of mine. The idea is to construct an artificial language
to exhibit semantic structure in much the same way as a
structural chemical formula exhibits the chemical structure of
a substance. The project has been examined by Professor
Carnap who found it to be "ingenious" . . .

But then look up: nothing except masses of bags stuffed full
of . . . see! *Here,* a second letter in a soap-powder box. From
the same man, dated eight months later:

Dear Mr. Norton,
I am very sorry that you did not reply to my last letter. I
suppose I must have offended you that I did not want you to
plagiarize my language idea. I should like to make it clear that I
did not for a moment think it likely that you would. It was just
that I have reached the age of thirty-four, and during that time
I have been diagnosed as suffering from schizophrenia, a form
of insanity . . .

Look up: supermarket bags, bags, bags sloshing off to the
horizon. And what's inside them? Sour-milk-colored objects.
Rammed inside every one of these plastic carriers,
stretching the entire length of the Excavation, rising here and
there into surges, filling tea-chest-sized cardboard boxes,
leaping up and taking over tables, seeping under doors,
splattering the insides of forlorn wardrobes and cooking
cupboards, submerging chairs, sloshing against the bed legs:
Bus timetables. Tens of thousands of them.
All of them out of date.

Time is very quiet in this house.
Nothing shifts in the potato light.

Not everything is disordered. Maps, filed edgeways on the mantelpiece, collapse from upright in order of grubbiness.

Several times a day, a car races up the road outside—an IT exec on his way to the business park, imagining he's found a shortcut past the traffic at the bottom of the hill. The noise simmers, boils, trumpets . . . crumbles back to silence. Minutes later, another heated noise—fuel injection, optical steering, scented airbag, blur of walnut dash—a different IT exec escaping the traffic at the top of the hill.

On stormy days, Simon's front patio kidnaps the wind. Billows of air kick up a panic, bang the windowpane, rattle yellowfly off the buddleia branches, and are beaten senseless against the coal-shed lock. The next day, resting under the ivy,

FRONT CAVE of Dr. Simon MINUS Norton's Excavation:
THE FICTION.

are jelly-baby packets, a shoelace, half a pair of spectacles, a bottle of Lucozade, half drunk, containing two cigarette butts.

The only regular noise inside the Excavation is from the boiler in the corridor across the room from where we're standing. Every now and then this ancient box of tubes gives a wearied huff of gas. In winter, the low whisper during the hours of 6 a.m. to 9 a.m. and 6 p.m. to 8 p.m. is like the hum of a mortuary fridge. Although we don't know it now, a bubble of carbon monoxide is building up in this corridor. A builder, who will appear in a few chapters' time, will discover this bubble with his electronic instruments. It is trembling disgustingly behind the door. If it weren't for the relieving swirls of fresh air from the top-floor tenants getting their bikes out of the corridor, it would long ago have oozed into Simon's front room and murdered the entire house. In a few months, gleaming new copper pipes will stream up and down the wall, spreading warmth, hot water and legal compliance.

Apart from Death and three bicycles, this corridor contains only one thing, tidily lined up along the shelves: gingham bags—the sort Chinese peasants carry when running away from floods.

Simon's basement feels like a resting place at the end of a long plunge.

I would have liked now to spend some time with you looking at the back room of the Excavation. It's tidier than the front. There's a large writing desk with three broken manual typewriters, and a mahogany occasional table—clean, free of dust—supporting a potted plant, now dead, its leaves the color of pie crust, and a snapshot of two children carrying a warthog. On one side of this room is a floor-to-ceiling bookcase on which everything is stored in marvelous order. To repeat: I would have liked to have investigated this, but—

I don't know if you've sensed it too—for the last few minutes I've been aware of a gentle extra odor of sardines coming over my right shoulder.

Someone is standing behind me.

5

You know, people think that mathematics is complicated. Mathematics is the simple bit. It's the stuff we can understand. It's . . . cats that are complicated. I mean, what is it in those little molecules and stuff that make one cat behave differently to another, or that make a cat? I mean, how do you define a cat? I've no *idea*.

Professor John Horton Conway,
Simon's former colleague

I don't mind cats, as long as they don't sit on my genitals.

Simon

"As I say . . ."

Simon's voice is monotonic. The equivalent of a glassy stare, for mouths.

"As I say . . ."

Simon often begins his sentences like this, "As I say," when he has never said anything of the sort before.

"As I say . . ."

If he's truly enraged at finding us down here, it will burst through eventually: a bubble in the mire.

"As I say, I am prepared to reconsider the matter of this book on the condition that my mother is the litmus paper."

Pushing the fish tin into his pocket, he yanks up his duffel, breaks away from two Marks and Spencer's bags oozing over his feet and barges toward the bed, shoelaces flapping. "My mother must be the test. You must write for her. If she approves the pages then they can go in the text." He extracts a book he's been carrying under his arm. "I have brought a thesaurus. Now, let's see: there are certain words I know she would prefer you not to use . . ."

The Dutch mahogany bed is rather high. He has to swing his duffel on first, reverse his bottom into position, take a breath and make a leap backward to get himself up onto the top surface.

Pressing the thesaurus onto the pillow with his fist, Simon peels it open in a way that makes me think of pastry dough and feel hungry.

"Would your mother like to hear you called 'unemployed'?" he says. "Unemployed, unemployed, unemployed . . ." dabbing his finger down the page. "Hnnnh, here it is: entry 266."

"But *I'm* not unemployed," I point out. "I have a job: I'm under contract to write about you. Do you have a job?"

"No."

"Then you are unemployed."

At mathematics conferences Simon is euphemistically listed as an "independent" researcher.

For the tax man, he turns the "un-" into a "self-."

When filling in survey forms, he puffs up his chest, rattles memories of past glory and describes himself as *ahem!* "In part-time work."

"The fact," he observes, "that the mathematics department here at Cambridge is not paying me doesn't mean I'm not *working* in the building anymore. I still have an office and 'independent researcher' is *not* a euphemism. It is a respectable designation, and does not mean 'unemployed.' Put yourself in your mother's shoes, then you'll understand. Would you want your children to think their father was a euphemism . . . ?"

My eyes return to his bag. It appears to be new. Every five or ten years Simon gets a fresh duffel and, for a few months, looks suspicious. The new fabric sparkles against his saggy trousers.

It's as if he's just passed a luggage shop and knocked off the first item he could reach in the window display, together with all its stuffing.

"Here we go: 'Unemployed, adjective: at rest, quiescent . . . motionless, stagnant . . . subsidence . . .' I certainly wouldn't like it if any of my children were written about like that. Hnnnh, let's see," he continues. "'. . . becalmed, at anchor, vegetating, deadness. . . .'" There's no stopping him.

He also disputes my use of "sacked."

"'Sacked' . . . let's see." He turns to another page. "'Let go'? 'Let fall'? 'Relinquish'? Aaah, '*liberate*'!"

"But you *were* sacked. You had a job, and you lost it because your students refused to come to your lectures and you were always sitting on a . . ."

"I was not sacked," he interrupts.

"According to my source, your students left in geometrical progression. First you had sixteen, then the next week, eight, then four, and when you got down to the last one, he died."

"I was not sacked," repeats Simon firmly. "I did not have my contract renewed. Everyone would agree there is a significant difference. And please do not say I was always sitting on a bus."

The most astounding mathematical prodigy of his generation did not get his contract renewed? A man who has the answer to the symmetries of the universe in his sights, dismissed like a Brighton coffee-shop waitress? "Sacked," I call it. "Sacked" in all but technical fuss.

But Mummy must not be told.

"I am not prepared to sacrifice her feelings to satisfy your artistic sensibilities," Simon sniffs. "The situation you are trying to manufacture reminds me of something I read in one of Hans Eysenck's popular psychology books. He describes a Victorian with the pseudonym Walter, the ambition of whose life seems to have been to have sex with as many females as he could."

"I *hardly* think . . ."

"Eysenck then expresses this point of view to put it up to ridicule: 'What do the feelings of all these females matter in comparison with the satisfaction of Walter's artistic needs?' As I say, my mother and children must be the test."

It is only now, recovered from the shock of Simon discovering me trespassing down here—a fact that he still appears not to have noticed—that I finally detect the flaw in his argument.

"But Simon, your mother died nine years ago."

"The principle is the same."

"And you don't have any children."

This is not the first time Simon's had cause to complain about my intrusions. When I was researching my first book ("which I think will also be your last") he made the mistake of popping his head round the door of my study while I was interviewing my then subject, and before Simon had a chance to scramble out of the room again, I'd snatched him into print.

"Twice winner of a Mathematics Olympiad Gold Medal . . ." I'd written as his footsteps fled, "my landlord is a generous, mild man, as brilliant as the sun, but a fraction odd."

"One fact to get right, and you got it wrong in four different ways," protests Simon now.

One: the International Mathematics Olympiad does not award medals or (mistake two) golds, it hands out numbers: 1, 2 and 3. Three: there is no such thing as a "winner" in these competitions: it is mathematics, not sprinting. You get a 1 for achieving a certain score or above. It is perfectly possible for all contenders to get a 1. Mistake four: three times—not twice—Simon scored this top grade, aged fifteen, sixteen and seventeen, and (although Simon insists he has forgotten this) one of those wins was with a triumphant 100 percent, a perfect flush, and another with 99 percent, one of the first boys in the

32

world ever to achieve this mind-frazzling triumph. Others have managed it since, but unlike Simon, they have had years of dedicated training, entirely skipped their adolescence, and looked like beaten-up tapeworms.

In just half a page of a biography *about someone else* I managed to misrepresent Simon in four ways, when all he'd done was have the bad luck to stray into my sight for five minutes.

"Four errors in half a page is, hnn, eight errors in a full page, which in a full-length publication such as you are threatening to make this one, comes to, aaah, 2,000 or 3,000 instances of disregard for fact. Oh dear!" he sighs. "Oh dear, oh dear."

True to expectation, the howlers in this manuscript have already arrived. "What do you mean," he says, submerging his arms into his duffel, for a moment looking puzzled, then following after with his head, as if his bag is eating him. "What do you mean"—he reappears with a clutch of papers, the first chapters, which I emailed him this morning—"that women have a habit of shrieking when they come across me?"

"Unexpectedly, when you're hovering next to my bathroom door. They do."

"It may have happened once," he permits. "I do not think that makes a habit. I do not think my mother . . ."

"Your *dead* mother, Simon. It's happened three times."

It's not his looks. It's the way he hovers outside the door, waxen and quiet. He's not there with any wicked purpose. He's been pacing up and down the front hall, tearing at his post or contemplating points of infinity in hyperbolic space, and just happens to have reached that end of the corridor when the bathroom door opens. His fixed stare gives him the impression of having enormous eyes. Muttonchop whiskers billow up the side of his face, as though his blank smile contained a fire.

Seated, second from right: A genius

DAILY MAIL JULY 4ᵗʰ 1967

PICTURE BY
ROGER BAMBER

THE BRITISH TEAM, from left, standing Cameron-Smith, Hill, Phair and Cullen. Sitting, Williamson, Garland, Norton and Davies.

BY GERARD KEMP

THESE eight boys are the finest young mathematicians in Britain. One of them is being hailed as a genius . . . a boy who is " head and shoulders above the other seven."

He is Simon Norton (seated second from the right), and he is only 15.

Master Norton, from Eton, scored 195 marks out of 200 to win this year's British Mathematical Olympiad for Schools. The second boy got 155.

The young Etonian and the seven runners-up from 10,000 competitors are representing Britain in the International Mathematical Olympiad, at Cetinje, Yugoslavia, this week.

They flew there last night and will sit their first examination paper tomorrow.

Some of the boys were apprehensive . . . they thought previous years papers were " tough." Not so Master Norton, who thought them " not too difficult."

Sample problem:

Clipping from the *Daily Mail*, found in a sorry state by the clothes closet, front room of the Excavation. Reconstructed by the biographer.

3. Given a right circular cone and the sphere inscribed in it. About this sphere is circumscribed a right circular cylinder, one of whose bases lies in the plane of the base of the given cone. Let v(1) denote the volume of the cone, and v(2) the volume of the cylinder.

a) Prove that the equality v(1) = v(2) cannot take place.

b) Indicate the minimum value k, for which the equality v(1) = kv(2) holds. For this case construct the angle at the vertex of an axial section of the cone.

(Bulgaria)

SAMPLE QUOTE from Master Norton:

Yes, I've seen papers from previous years and I must say they don't seem too difficult.

34

Sprouting under his nostrils is half an inch of bristle where his electric shaving machine—based on circular movement—doesn't reach into the corners of his nose. His stillness suggests someone plotting ambushes on a safari, or one of those people who squat in ponds with weeds on their heads, shooting ducks.

The woman shrieks. Mid-shriek, Simon does nothing, as though he's thumbtacked between two seconds. Only once the screams have died down into gurgles of relief and apologies does he shake himself free with a heave of breath.

"Hnnn!" he says.

"Hnnnn," he repeats.

Relieved to have resolved the situation so deftly, he thumps downstairs to the Excavation.

Another error he has noticed in these sample pages: why do I say he smells of sardines in tomato sauce? They're *not* sardines, they're kippers. They may, on occasion, be mackerel. And in all cases he buys them in oil. He dislikes tomato sauce.

"If I can't say you smell of sardines in tomatoes," I retort, "can I say you smell of fatty headless fish?"

It's essential to emphasize that in no sense of the term is Simon mad. He's covered in facial hair and wears rotten shoes and trousers for the opposite reason: too much mental order. He burps; he makes elephant yawns without putting his hand over his mouth; he thinks you won't mind knowing about the progress of his digestion; and he goes on long, sweaty walks, then doesn't change his clothes for a week. But what else can he do? Everybody is messy somehow, and there's no other place for Simon to store his quota. Inside his head there's no room: all the mess has been swept out. It's as pristine in there as a surgeon's operating theater.

Another word he doesn't like in my manuscript is "stomps."

"What do you mean, I 'stomp'? How do you know I 'stomp'? I don't believe you can hear me from upstairs. You're not suggesting I 'stomp' on the ceiling, are you?"

As for my description of his floor . . . "Oh dear," he groans, conclusively.

Suddenly, Simon loses interest. Although his face has no time for expressions, his legs and arms want to get on with it. He starts to wiggle his hands; his head begins to rotate; then, without explanation, he drops the thesaurus on his bedcover, bolts from the bed, dodging a wave of Asda bags ("Sainsbury's, Alex. I find it enhances one's appreciation of a book if the facts are correct") and hurries to the kitchen, gripping his peck of peppered kippers. Through the connecting door, I watch his hair weave around the lightbulb like a gray feather-duster. A large, disjointed man, he can move with surprising litheness.

People such as Simon—unknown, living people—don't trust words. Words may be a familiar method of communication (although Simon generally prefers grunts or showing off bus tickets), but that doesn't mean it's respectable to make a living out of them, especially if you're a sloppy scribbler with a light-hearted attitude to truth like me. Words are too nuanced and potentially destructive to be left in the hands of someone so unrigorous. A straightforward four-letter noun beginning with f—

"No!"

—defining your style of accommodation, and bang! The entire disciplinary force of Cambridge City Council rushes up the hill with clipboards to snap, tick and bylaw you into a magistrate's court.

For Simon, the world is a leaky place. You have constantly to be on your guard against the seeping away or sudden disappearance of comfort. He imagines that this book ("if it ever comes out") will be "bedside reading" for housing inspectors. He thinks they might run him out of the city.

If that's what words can do when wrongly applied to a few cubic feet of basement air enclosed by bricks and bramble-bush-covered windows, what massacre will they perform on the central object of a full-length biography—which is a trillion misunderstandings-in-waiting—i.e., a living human being?

Simon says he doesn't stomp. I say he does. Simon says he should know, since a) he does the non-stomping and b) he's closer to his feet than I am.

"But if you are stomping on the ceiling, then my ears are closer," I observe. "Biography—especially biography of an unknown person—is not and cannot be about reality." I follow after him to the kitchen. "It's no more about reality than, say, say . . . minus numbers. And just as the solution to the problem of the impossible existence of minus numbers is to realize that they are not real things at all, but something you've done to positive numbers, i.e., you've 'minus-ed' them—in short, minus numbers are verbs, not nouns—so in biography, it's not the real subject, but the active, i.e., verbal, relationship between the biographer and subject that . . ."

"Mathematicians do not think of negative numbers like that," interrupts Simon, tugging at the mackerel tin, which has somehow got wedged in his pocket. "We think of them as real objects. Exactly as real as positive numbers."

"The reason that a biography of an unknown person cannot be about reality," I continue regardless, "is because the reader will fall asleep. Reality is too bland. An ordinary person doesn't have the dramatic and universally appreciated facts of the famous to rely on. They've only got the oddness and power of their character. So," I say, expanding my chest with the sudden conviction that I am going to be able to complete these sentences neatly, "a biography of the unknown has to be a biographer's effort to *interpret* facts, his impression of the facts—what has been done to the facts by his brain. It's about one person's mumbled attempts vaguely to interpret what they dimly think they might have seen on a misty day in another person's possible behavior, but which they quite possibly haven't; and any biographer who puts pen to paper claiming his motives are objectivity and truth is a fraud. Biography is not mathematics. It is not bus timetables. What matters is not whether or not you 'stomp,' in fact, since who can know that as a fact, but that I *think* you stomp, and by the way, aren't you supposed to take the sweet corn out of the supermarket bag before you put the tin in boiling water?"

Squeak of a tap; the cymbal clatter of high-pressure liquid on thin cooking steel; the castanets and maracas of bubbles; muffled turbulence as the pot fills.

Simon's wolfish. While we were trespassing through the rubbish in this basement, he'd been on a moonlit hike around the city distributing anti-car newsletters for an environmental campaign group called Transport 2000, and it's emptied his belly. After buses and trains, the thing that matters most to him is his digestive tract.

Small headless fish are his favorite food. Except when in Montreal, Simon boils his kippers in the tin, "to save on washing up." Kippers come in a different-sized tin in Canada, and "I don't want to take the chance of doing something wrong." In Montreal he eats frozen fish in supermarket display packs—not because he prefers it but because the label tells him what to do, which is comforting, although he never grills, "because you can't see what's going on."

Like Ludwig Wittgenstein, Simon does not enjoy variety in food.

"I like to find a formula that works and stick to it," he insists, stepping out of the kitchen to make sure I understand.

"I once found myself in possession of mackerel in curry sauce because I'd failed to look carefully enough when in the supermarket. I couldn't finish it.

"Yes, I am a worrier. My mother was a worrier."

Simon is incapable of frowning; his expressions are limited to petulance, grinning and vacuity. He adopts the last, and returns to the stove.

Mackerel Norton, the dish he is preparing this evening, is his Number One meal. It comes in two forms: finger-scalding hot and body temperature. Tonight, he's having it hot.

Mackerel Norton
for one

1 tin of mackerel fillets, any sort, as long as *not in tomato sauce.*
1 flavored Batchelors Chinese packet rice. ("I sometimes use 'Golden Vegetable.'")
2 pans of boiling water.

Put first two in the third. Bubble rice frothily for correct time. Release rice, spurt open mackerel, eat on bed with much hand waving and gulps of cool air.

He would, if he could, eat Mackerel Norton seven days a week; but world events and the pressures of anti-car campaigning are such that he can barely manage to get it three days in a row. The rest of the time he gobbles two forms of takeaway (chicken biryani and chicken in black-bean sauce), chili-flavor crisps from Morrisons and Bombay mix, a spiced Indian snack.

This evening Simon has accidentally picked up a different-flavor packet rice, and is alarmed. Cooking instructions are suspect to Simon. They are the route errors use when they want to sneak into your stomach. Why should one flavor respond to hot water in the same way as another? How can you be sure that one rice packet, representing the products of a country containing yellow people in blue boilersuits, should be treated the same as another packet, from a country 16,000 miles distant from the first, with brown people and cactuses? Cooking instructions have no appreciation of the slyness of variables.

"Uugggh, do Mexican vegetables boil in the same way as Chinese?" Simon asks, waving the packet at me through the kitchen doorway.

In Simon's kitchen there are no cobwebs. An aerosol of grease has killed them off. If you stand on a footstool, it is possible to find—original inhabitants, from before the Extinction Event (Simon's purchase of the house in 1981)—dead spiders inhumed above the wall cupboards, in the Cretaceous layers of fat.

There is evidence of urgent eating everywhere. The oil slicks on the melamine surfaces; eyebrow hairs embedded around the sink; foot and shoe grime that has gathered on the plastic embossed-tile flooring, making it look almost as though there is a rug on top; the curtains of grease moving down the sides of the sink like textured glass.

Simon is not unhealthy. The principal source of serious infection in any house—the water supply—is cleaner here than in most places, because the attic in which the water head is stored is used as a room for tenants, and is therefore easily accessible and frequently checked. I can vouch for the fact that there are no mice floating in it, or spiders, woodlice, bloated and putrefying snails, or dead rats, as there certainly will be in the water tanks belonging to some of the people reading this sentence.

He is not unhygienic, except in the eyes of today's dainty obsessives and kitchen-product advertisers. He has a bath once a week and cleans his teeth daily. But he is not frightened of his digestion. Simon's connection with decomposing food begins and ends, openly and honestly, as it does with all animals at ease: with a squelchy chew at one end and a sigh of release at the other.

In a tidy kitchen, every knife, plate, whisk, frying pan, coffee mug, ladle, tea strainer, and chopping board and all machines are stagnant with cleanliness, with the exception of the dishwasher murmuring disinfectant-speak under the sink. The object of the tidy and twee housekeeper is to remove all proof that he is a functioning organism.

In Simon's kitchen, Hunger has slobbered everywhere.

Yellow smears splashed along the left-hand worktop are from cartons of chicken biryani, the lids ripped off; the drips of purple, slightly granulated, are Ferns' brinjal pickle; the intermingled slops of ochre green, Mr. Patak's mixed pickle.

"And what's wrong with Mr. Fern's mixed pickle?"

"I don't know. I've never tried it. Do they do one?"

Both types of stain are the result of scraping contents from jars with a plastic spoon that is too short, and rushing the findings back on a bombing run across the sideboards to the now dead chicken. By the sink, chicken in black-bean sauce has added a brown tinge.

This rancid atmosphere and the cold, soporific mood of the main rooms, together with the almost undetectable whiff of furniture polish from the paintings and mahogany items that Simon inherited following the death of his father— a homeopathic dose of plushness—combine to give the Excavation a pleasant smell. Warmed up, with perhaps a squeeze of lemon and lime shaving cream thrown in to suggest Life, it might even be cozy.

All the same, it's easy to get carried away by this bomb site. Simon isn't universally messy, even outside his head. He's as fussy as a surgeon when it comes to planning a journey. He manages two homes (he has a flat in London), has a turnover of satisfied tenants, and is never behind with bills, legal documents or financial dealings with his accountant. None of these is true of me. In addition, his transport newsletter comes out once every three or four months; is twelve to sixteen pages

A list, from 1992, of a few of the bus and train journeys Simon made that year. He has a pile of such lists, over a foot high.

long; single-spaced; eight-point type; covers hundreds if not thousands of unfailingly accurate details about new routes, closures and timetables; and keeps careful account of all local outrages by the government Highways Agency. When Simon wants there to be order, he's unmatchable. When not, a colostomy bag is not more disgusting. Simon insists that this basement is his catalogue: all it needs is pruning, sorting out, filing, and it will be an invaluable library of documentation.

"A documentation of what, exactly?" I ask while he sits down to his supper.

"Where I've been?" he suggests.

I call it his middenheap. These papers are just bones: all that is left after Simon's banquet on their information relating to buses and trains: the public-transport detritus of a monstrous feast on facts that began when he was three.

"How about if I take the focus of the story off the floor and into the air?" I suggest breezily, returning to the battle. "'One of the greatest mathematical geniuses of the twentieth century lives beneath my floorboards,' I could begin, 'in the dank, fetid gloom of his subterranean . . .'"

"No."

"Not dank?"

"No."

"Or fetid?"

He shakes his head.

"How about miasmic? I quite like miasmic. It sounds poetic."

Also no good: "Ungh-ungh."

I take a deep breath, slowly let out air and reach across for Simon's thesaurus. "Ponging?"

6 The Monster

That's the name of Simon's special area in mathematics, because of its gargantuan complexity and fiery insight into the fundamental structure of our universe.

No one knows what the Monster looks like. It can be detected only through its mathematical traces. Like shadows and ghosts, it inhabits a penumbral landscape between abstraction and solidity.

The Monster belongs to an area of mathematics known as Group Theory, or the study of symmetry.

Groups are represented in textbooks by tiresome grids of numbers similar to sudoku tables, yet they are among the most startling investigative tools in human thought. Quantum Theory, Relativity Theory, predictions about the number and types of sub-atomic particles, the codes used to scramble military and financial information—all of it fundamentally reliant on the study of Groups. They have even been used to investigate incest among Aboriginal tribes.

A sudoku table has nine rows and nine columns of numbers.

The Monster has 80801742479451287588645990496171075700575 4368000000000.

*7

Introducing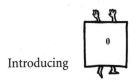

To understand Simon's particular genius—how it developed and why for a few years he led the braying pack of mathematicians hunting down the Monster—the reader needs to know about squares.

On the face of it, the study of symmetries is a subject for children. A square has symmetry: you can rotate it, and the result looks just as if you'd done nothing at all:

The same goes for an equal-sided triangle:

A circle, cube, sphere and a host of other shapes with names like dodecadodecahedron (twenty-four faces) and icosidodecadodecahedron (forty-four faces) each have similar symmetrical properties.

In order to develop mathematics out of such simple stuff, we have to keep a diary of these symmetries.

For example, to keep track of these four moves, we can represent them like this:

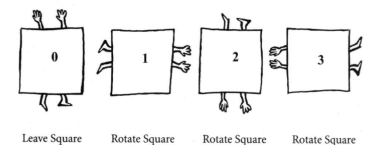

| Leave Square alone | Rotate Square by one turn | Rotate Square by two turns | Rotate Square by three turns |

Note that there's a sense of self-containment about this set of operations. A square has four sides and therefore only four distinct ways of rotating. After that, you've exhausted all the possibilities. No amount of rotating will paint it green or puff it up to twice its original size. Other operations are needed to perform that sort of thing.

If we rotate a square in any of the above four ways, it still looks to the outsider just like the square we started with:

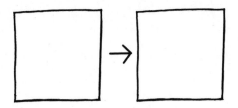

But, privately, we know we've been fiddling. For example, if we rotate a square through two turns (i.e., flip it head over heels), we can represent this:

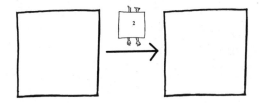

In other words,

signifies the *act* of swiveling a square through two 90-degree turns, without anybody noticing.

Naturally, if you turn a square by one turn through 90 degrees, then do it again, that's the equivalent of two 90-degree turns overall:

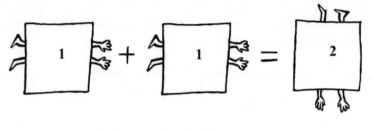

$$1 + 1 = 2$$

Similarly, rotate a square once, followed by two more turns, and the result is equivalent to three turns. You've almost gone the whole way round:

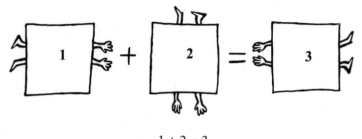

$$1 + 2 = 3$$

And so on. Rotate a square by two turns, then do nothing, go off and play with somebody else's crayons, and no one's going to be fooled—it's still just two turns:

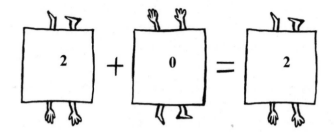

$$2 + 0 = 2$$

A square looks just the same after any combination of these operations, or all of them:

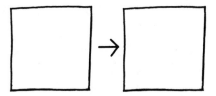

The figures with arms and legs are simply diary entries to keep track of the secret things we've been doing to the square in the playpen.

What happens if we turn a square, say, five times? That's the equivalent of spinning it through a full cycle, then throwing in an extra single turn for good measure:

Group Theory isn't interested in recording such clever-clogs stuff. Turn a square round five times and you might as well just have turned it once. It's the final outcome *only* that matters, so it's put down as an ordinary single turn:

So, although it seems possible that:

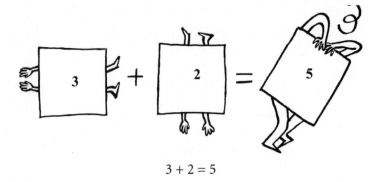

$$3 + 2 = 5$$

because the first four turns make a *complete* rotation, head over heels, back *exactly* to where we started, we ignore them as wasted effort, and just focus on the one leftover turn, which got us somewhere:

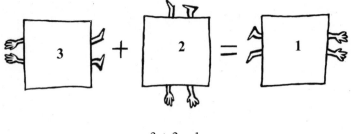

$$3 + 2 = 1$$

In this respect, rotating a square is the same as rotating the hour hand on an ordinary clockface. If it's two o'clock and we add twelve hours, we don't say it's fourteen o'clock (unless we're being tiresome). We say it's two o'clock again.

All these combinations of turns can now be written down as a table. These are the lifeblood of Group Theory. Every mystery of this secretive, sly subject is contained in such tables. The one that applies to turns of a square is:

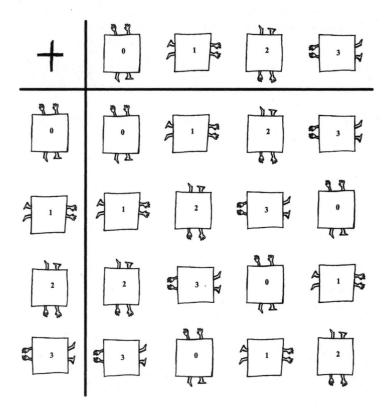

It's read in the same way as the distance chart at the front of a road atlas. It's not a calculating device; it's a secretarial way of keeping track of information. If you're six years old and want to remind yourself what happens when you turn a square once, then turn it again, i.e.:

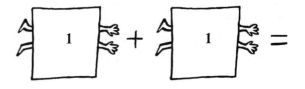

take the row corresponding to 1, run your chocolaty finger along until you come to the column corresponding to 1, and there's the answer—it's equivalent to two turns:

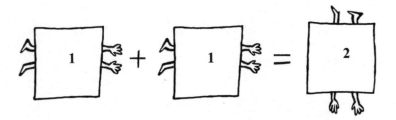

What's baffling is that there can be anything complicated enough about this "study" of symmetries to bring it out of the playroom in the first place.

45

The papers that slosh about in the basement are (Simon insists I am to say) "carefully, hnnnh, arranged and, uuuh, being sorted in plastic bags."

But I've put my foot down over this: "That's a lie, Simon. You're telling me to make things up."

"I've not noticed your reluctance on that front before."

Simon was taught mathematics by the number 45.

The first written evidence we have of 45's significance in his life comes from an Atlantic blue notebook dated January 1956 (a year before Simon started school) and titled

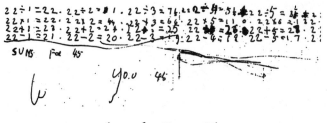

Inside, Simon addresses mathematical problems to this number:

(sums for 45, you 45)

performs amusing numerical games:

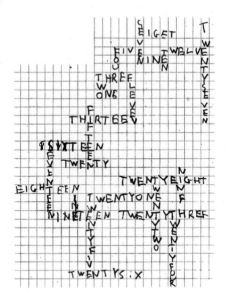

and emerges briefly, porpoise-like, from his researches to write letters to her:

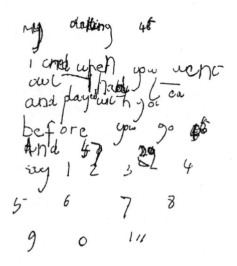

before re-submerging in a glug of numbers.

Sometimes, 45 wrote back:

My darling Simon
I am sorry I
had to go out this afternoon
and couldn't play with you
after your rest. You shouldn't
have cried. Please don't cry
the next time I go out. I shall
always try to be home by 5 o'clock
so that I can play with you
before you go to have your
bath. Lots of love and kisses
from
Mummy.

45   2 15   1/1

45 was Simon's number for his mother. She was the one who taught him math, up to quadratic equations. Astounding, for a British housewife in the 1950s—no one in the family can explain it.

Once the first Atlantic blue notebook was finished, Simon and his mother started another, in February 1956, just before his fourth birthday. Her handwriting is on the cover this time, and she calls him a monkey:

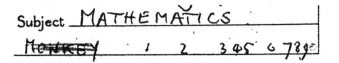

Subject MATHEMATICS
MONKEY   1   2   3 45 6 789°

I wonder who crossed this fondness out.

Once, during these preschool years, Simon was distracted by a word:

But he soon snapped back to numbers with gusto:

9 eSimon

Simon has five types of grunt:

Ask him about his mother,

his father,

or his parents' marriage,

and you get a happy grunt for the
first, a puzzled one for the second, and an incomprehensible
bleat concerning the photo above.

It's possible to extract more interpretation, if you work at it.

"Simon, what do you mean, 'Uuug*heugh* . . . gghuaha . . .
ehh*ghH*?'"

"Haaaghuggh . . . *oooh* . . . ghughghEH."

"These are your parents. You grew up with them. Is this
photo an accurate reflection of their relationship or not?"

"Aaaghurghh . . . gghahuugh . . . eeehghuGH . . . is that a
dog in the middle? Oh, it's a bag. If it had been a dog these
might not have been my parents, just people who looked like
my parents, because we didn't have a dog."

"Don't you think there's something remarkable about this
picture? Sitting back-to-back on a pebble beach in Southern
England with something that looks like a dead bulldog
between you?"

Simon looks trapped and panicky.

"Or is this just a snapshot of what marriage always is, to you?"

The grunt-bleat means bafflement. Why should he find something odd about the photograph? It's a photograph: one of those things (so rare in life) in which a fact is made immutable. Why muddy it with interpretation? Why, if you do muddy it, pick on that *particular* interpretation? It could be one of a million others.

Simon is, verbally, one of the most adept and playful people I know—as long as he doesn't have to speak.

Or use metaphor.

Or comment on photographs.

Simon's parents' marriage (according to other members of the family) was not happy or unhappy, just mannered and soulless.

The puzzled father-grunt Simon gives about his father signifies a lack of interest. The happy mother-grunt, "loveliness." Loveliness is the only adjective Simon associates with Helene Norton. She "embodied" the word, he says. There isn't need for others—and he doesn't mean "loveliness" because of her startling beauty, which Simon claims he'd never noticed until I started ogling her, but "loveliness" because of . . . because of . . . uuugggh*hhhAH*! Grunt Number Four: Frustrated Grunt.

What's the point of me demanding new words when he's already given me the one that works to perfection?

"Loveliness" does not mean uncensoriousness, however. When his mother was alive, Simon used to visit her in London every two or three weeks, but she refused to greet him until he'd had a bath.

Then she would criticize his clothes.

"If it wasn't one thing, it was another," remembers Simon forlornly. "I got the feeling I could never satisfy her. Did it count if your clothes were wrong in the period *before* you'd had a chance to spruce up between the front door and the bath?"

After trying to "spruce up" he'd step out of the bathroom in the fresh clothes that his mother kept in a cupboard, ready for his visits, and expecting now to be allowed to kiss her hello. "But I'd almost certainly forget something, and she'd draw attention to that one thing and home in on it. My shirt was wrong, or my shoelaces were undone. I hadn't done up my trousers correctly."

He gives a purgatorial groan.

"It was too much for me."

Two or three days after his mother's death (Simon remembers it as "rather too quickly") he and his two brothers let themselves into her five-bedroom apartment near Baker Street, and began picking over her possessions. From their mother's cupboards and drawers they extracted everything small and unbreakable and piled it on the floor. Then they shuffled among the piles—in my image of this spectral, sacrificial scene I imagine them as three tall birds, and hear the clicking of their feet on the parquet floors—plucking up anything that took their fancy.

There was only one item Simon wanted: a photograph of his mother in old age.

There were paintings of her, when she was young and glorious. Simon wasn't interested. He'd had nothing to do with her in those days. He doesn't like portraits at the best of times, but he prefers at least that they correlate to an image already in his brain.

He held out his arms, eyes closed, to any other things the brothers didn't take, then brought the fifty- or sixty-item windfall back to the small flat he owns in London. There he laid them out, ten layers deep along one edge of the living room, like drying fish fillets.

Simon tells me he would like to hang the pictures up.

His mother has been dead nine years now, but the haul remains stacked against the wall, curing itself slowly of connotations. "Loveliness" now resides only in the photograph of her old age and his memories.

The leftovers in his mother's apartment—her letters, wrapped in pink ribbon, from a man who was not Simon's father; her skirts and chemises, brooches, diamond pins, fur coats, perfumes, old swing-band records—the brothers sold, gave away or threw in the dustbin.

"But you also got all your old school reports and exercise books, and the folder of newspaper clippings about when you won the Maths Olympiads and went to Cambridge, your IQ report?"

"As I say, I didn't want them."

"Then why take them?"

"Why *not* take them?"

Simon is always eager to drop in schoolboyish retorts like this. The trick is to become instantly absurd.

"Would you have taken them had they been roast chickens?"

"Heh, heh, heh, hnnn. I took them because it's the sort of thing people do take, isn't it?"

See? Simple, when you know how.

To Simon, correct conduct is like a wood. It has many trees, which represent how things ought to be done; one tree for each circumstance. It is a large wood, sterile and rather dark. The stormy forest where he goes to hunt for the Monster is infinitely more comforting.

Here's Simon's brother!

Hello, Michael!

He doesn't have much to say.

"Is it surprising?" he protests, leaping up, holding out his hand—a strong shake. "I'm ten years older than Simon is. We were like different families. I studied chemistry at university, not mathematics; that's a different language. Simon is interested in chemistry also? Really? I never knew. His favorite element is boron? I'm surprised! Would you like some tea? Organic Lapsang or elderflower?"

Michael Norton OBE is the author of *Writing Better Fundraising Applications, The Worldwide Fundraiser's Handbook, The Complete Fundraising Handbook* and *Getting Started in Fundraising.* Money—in particular other people's money—is a big subject in Michael's life. He wants it to pay for environmental revolution.

His latest book is *365 Ways to Change the World: How to Make a Difference—One Day at a Time.* Each day of the year is allocated a noble deed:

January 5: "Start drinking." Reduce "beer miles" by giving up sewage brands like Heineken or Budweiser, and brew your own beer using oysters and wild rice.

February 22: "Say no to plastic bags." There are now 46,000 pieces of plastic waste in every square mile of the world's oceans. In Australia, eighty million plastic bags are added every year to the mist of garbage that floats across the scrub there. Cows eat them and die; then the sack re-emerges from the rotting flesh, is cleaned off by rain, scooped back up by wind, and bundled along to be eaten by another cow: it is, biologically speaking, a protovirus. Simon is therefore a force for salvation. He keeps these protoviruses out of reach. If it weren't for him, thousands of extra plastic bags from the Excavation would be tumbling through our fields and woodlands.

May 1: "Join the sex workers' union." Fight to give prostitutes access to health care, safe places to work and legal support against rapists and pervy Italian prime ministers. "Membership is free."

"Michael Norton is a one-man 'ideas factory,'" bellows the *Guardian.*

"You know, he knows when he comes to dinner here dressed in a dirty T-shirt that he's doing wrong," says Michael, stooping under the lintel of his cottage door (he lives in Hampstead, but the house looks as if it's been airlifted from beside a village brook in Hampshire) and balancing a tea tray. "But there's no point telling him. You'd physically have to burn his old clothes before he'd get rid of them."

He's brought a photograph album into the garden along with the Victoria sponge cake. The album's green, with a cushioned cover, from 1954, and it's all the paperwork Michael has that includes Simon. They are not a sentimental family.

"That's my hand at the edge there, sorting out his food, even at that age. We're at our summer bungalow in Ferring. This is David . . ." David is a friend who later murdered his wife by bludgeoning her to death with a champagne bottle:

"And here's Simon aged . . . oh dear, not a pleasant-looking young man":

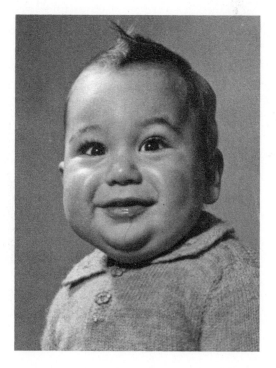

"Our mother doted on Simon. She was really proud that he was a genius. I don't think she ever understood why he didn't sustain that. I mean, he sustained it in his brain, but why is he not a professor? Why has he not got a proper job?"

"He's too peculiar," I suggest.

"He's not that peculiar," retorts Michael sharply, catching me out, correctly, in one of the phrases I have lately come to use about Simon without thinking. "There are lots of peculiar people in Cambridge. Half the lecturers that I had as a student there were peculiar. There must be somewhere that would give him a home."

He taps his china cup of elderflower irritably.

"All I can say is that since our mother died, Simon's become a different person. I noticed that almost immediately. He's got more sociable. When he comes to dinner, he's much more at ease. Instead of sitting in a corner reading a book, as he did when she was alive, he joins in. I've bought him three clean shirts which hang in a wardrobe here, for him to pick up whenever he comes to London.

"I think my biggest triumph is persuading him to get rid of his money. Did you know he gives £10,000 a year to campaign against cars?"

Francis Norton, Simon's middle brother, works here . . .

. . . in a jewelry shop.

Francis brings in the family money. The company, founded by Simon's great-grandfather, is the oldest family-run antique jewelry business in the world, patronized by the Queen, pop stars, fringe aristocracy, footballers (if they know what they're about) and all London people with 100-acre second homes in Wiltshire.

Ten years before Simon was born, S. J. Phillips established itself as the epitome of Englishness by taking part in a famous wartime deception called Operation Mincemeat, later dramatized in the film *The Man Who Never Was.*

In April 1943, a Spanish fisherman discovered the decomposing corpse of a man floating off the coast of Andalusia. Documents on the body identified him as Major (Acting) William Martin of the Royal Marines. He was handcuffed to a briefcase, which contained a bunch of keys, an expired military pass, two passionate love letters, a picture of a woman in a swimming costume ("Bill darling, don't let them send you off into the blue the horrible way they do"), a £53 bill for an engagement ring and a letter from Lord Louis Mountbatten to General Alexander revealing the plans for the Allied invasion of Europe.

Spain, though neutral, supported a very efficient network of German agents. They soon found out about the drowned Marine, got hold of the briefcase and carefully extracted the top-secret letter from its envelope. The British had made the greatest intelligence blunder of the Second World War. With ample time to prepare his defenses, Hitler now knew that the Allies were going to invade Europe through Sardinia and the Peloponnesus: the Germans transferred the 1st Panzer Division to Greece and started laying minefields.

It wasn't until the British got the corpse of Major (Acting) Martin back, a fortnight later, that they knew the Nazis had definitely fallen for the trick. British Intelligence had folded the fake letter to Mountbatten only once before sliding Martin's dead body into the sea from a submarine off the coast. When the body and effects were returned, investigators spotted under

---

Excerpt from an interview with Francis and his wife, Amanda

Amanda: When I met the Norton family, I thought they were all so bizarre.

Francis (nodding): My mother was very, very old-fashioned. It was, you know, "It's four o'clock in the afternoon, you can go see your mother." We absolutely adored her.

Amanda: They were just so Victorian. I'd never met anything like it in my life.

Francis: My father always said his ideal was to have a tail-coated butler behind every chair.

Amanda: I've never known parents who were so unphysical. In the morning Helene, their mother, would go off and do her charity work, then come home and have a long cigarette, put on her caftan for the afternoon, and sit there doing the crossword puzzle.

Francis: Terribly unfulfilled.

Amanda: I remember once—this is how old-fashioned they were—I'd just had my son Alexander, he was about nine months old, and Dick [Simon and Francis's father] was standing there, with the table laid with all the silver, and Domingo the butler hovering around. And Dick looks at me and says, "Amanda, darling, has Alexander started masticating yet?"

a microscope that the letter had been carefully refolded, creating a second crease.

A month later, Britain and America began their assault 300 miles west of the location indicated in the letter, through Sicily.

S. J. Phillips, Simon's family firm, provided the £53 engagement-ring bill—it was seen as the touch that the Germans would regard as unfakably, quintessentially English.

Francis is Simon's savior. It's because Francis keeps the family firm alive and profitable that Simon has never had to have a job or a mortgage and, despite using seventeen different variants of bus, train and visitor-attraction discount cards, doesn't actually need a single one of them.

A mild, self-effacing, apparently undogmatic man (I've met him only twice), Francis lives on the other side of Hampstead from brother Michael, and has the talcum-powder-dusted look of the very rich. He is an accomplished cellist.

Every year Francis or Michael invite Simon to their house for Passover; and every year Simon arrives with his shoelaces flapping, his duffel bulging, his bus timetables and his smells, and eats all the parsley.

Now, back to grunts.

The fifth type of grunt emitted by Simon is metaphorical—it's not a guttural sound, it's a full sentence. The mathematician Professor John Conway calls it a "Thank you, Simon," defined as "a statement that is indubitably true, but the relevance of which is obscure." For example, in the middle of a discussion with me about whether the Monster might in fact not be a large object at all but something very small and everywhere, like a flea, Simon will burst out:

"Incidentally, I was once going on a train and the conductor pronounced that we were now approaching 'Manea, the center of the universe.'"

What can you say after he's said that? What does he mean? That the flea-Monster, which Simon suspects contains the solution to the symmetry of the universe, is living in Manea, a village in the Cambridgeshire Fens? It can't be that. Simon is not a lunatic. Maybe it's just the word, "universe." But he clearly expects some sort of reply. So, after a suitable pause, you murmur, with a slight doff of the head,

"Thank you, Simon."

Then you attempt to pick up the pieces of the shattered conversation.

It's important to realize that this fifth type of grunt never comes about because Simon isn't able to keep up with the discussion, or because his brain has short-circuited and popped out the non sequitur in a fizz of misfired neurons. They appear for the opposite reason: he has dashed too far ahead, gone off on a side path, left the ponderous, sequential-talking rest of us behind, raced up into the hills of puns and synonyms and humorous, *leapingly* interconnected memories . . . then jumped back with the result, waving his arms and grinning in triumph, like a child ambushing us from behind a tree.

As the fact dawns on Simon that no one has the foggiest idea what he's talking about, he is not resentful. Politely, he allows the intensity of his grin to slip away. Measuredly, he rejoins the conversation.

(1) Happy, (2) puzzled, (3) incomprehensible, (4) frustrated, (5) phrasal ("Thank you, Simons"). Sometimes, Simon will go for weeks without offering anything to his biographer but one of these five grunts.

And then, *PING!*

An email comes.

And behind the grunts, a man.

Monday, February 8th, 9.19pm
I'm sorry, I can't make head or tail of the last chapter you sent
me. I think that any reader who shares my way of thinking will
be completely bewildered.

Tuesday, February 9th, 1.17am
I can't follow the thread of your writing. If I were someone I
didn't know rather than myself, I suspect that in reading it I
would have problems following the story even if I could
understand the sentences. Incidentally, this is not something
I'd say with your previous book. There I could understand
your description of Stuart, my problem was you gave me no
motivation to understand his character.

Wednesday, February 10th, 12.32am
I'm not sure what you mean [in Chapter 10] by my "jocular"
attitude to mathematics, but never mind. You've got the
calculations wrong—2^8 is 256, not 4096, which is 2^{12}, and the
others are similarly shifted. I don't understand your bit about
numbers floating in the sky. No, I haven't a clue whether it
was a right or left leg that the duck was missing.

[This is followed by fifty-two lines explaining the story of the
legless duck, which also includes a self-playing piano, an
inferno in the Channel Tunnel, admission that he reads a
magazine called *Cruising Monthly*, and a threat of
imprisonment by gas inspectors.]

Wednesday, February 10th, 12.48am
I know what the word "jocular" means. What I don't know
is what you mean when you describe my love of
mathematics as jocular. I might be jocular, but how can my
love of mathematics be? I don't know what you mean by a
"Rabelaisian" series (and don't say "in the style of

Rabelais"!). However, unlike you I do know how to spell the word.

eSimon, the Simon who logs on to his computer at one in the morning, is a different man from Simon the grunter: eloquent, fluent, conversational, reflective, poignant, sometimes funny and—if the subject matter has anything to do with my attempts to understand genius, popularize mathematics or write biography—acerbic.

Simon's interview with Kevin,
resident of Cambourne, in 2016

(An example of Simon's clear, fluent, amusing writing style. Abridged from an editorial [2006] in his Public Transport Newsletter, which he writes and publishes three or four times a year.)

Q: *What decided you to move to Cambourne* (a village outside Cambridge)?

A: We chose Cambourne because there was a direct bus link to my job in Papworth, and we could also get buses to St. Neots for trains to London. It also seemed a good place for my wife's aging parents. And we hoped our house would be a good investment—its value would have gone up had the east–west rail link been built close to the A428, as recommended by the London–South Midlands Multi-Modal Study.

Q: *But I gather things then went sour.*

A: Yes. In 2005 the bus links to Papworth and St. Neots were reduced, and I found I had to cycle in most days. The main road was very unpleasant, and the side route via Elsworth took twice as long. Then in 2006 came the Council's budget cuts to buses. In 2007 the A428 dual carriageway opened, our road became an "overspill A14," and Madingley Road became clogged, making our buses increasingly erratic.

Q: *But things are better now, aren't they?*

A: Yes, in one sense. The big stores left the city center because they realized people didn't want to have to put up with gridlock every time they went shopping. But there's the downside that it's now much harder to get to the shops by public transport. Nor could we use Internet shopping as there was rarely anyone in the house to accept deliveries, apart from my mother-in-law, who was often asleep, and even when she was awake she could never get to the door on time.

Q: *How has your family been coping?*
A: My father-in-law died in the bird-flu epidemic. My mother-in-law has become increasingly frail. Visiting the children in Oxford and London is a problem—the bus to Oxford has been taking ever longer because of growing congestion, and it's a long walk from the city center to the rail station. For a time we tried the coach, but then they moved the coach terminal to the rail station too . . .

Q: *Have you ever thought of buying a car?*
A: Yes, often. But then we'd ask, how could we face our children knowing we'd helped to ruin the world for them? Our generation has badly betrayed our children's.

Q: *I gather you're leaving Cambourne soon?*
A: Yes, we'll move to London or Oxford as soon as we've settled on a place for my mother-in-law. Good riddance—to the Cambridge area I mean, of course!

See *www.cambsbettertransport.org.uk/newsletter93.html for the full version.*

10 Mars

People do sometimes tell me how nice I am looking (e.g., at my mother's funeral) when I wear new clothes, but it always makes me feel very embarrassed. I say, "I don't want to know that." I don't want to be thought of as someone for whom personal appearance is important.

Simon

"I'm going to see a Martian. He aaah, hnnn . . . *it* lives in Woking."

Simon blocked my sun, his duffel swinging slowly to a stop after his unexpected rush off the pavement at my café table.

"Hnnn, aaah, uugh. As I say, my grandmother lived where the Martian is. Hnnnh. Would you like to come too?" Spring on earth! Simon giving me encouragement!

I jumped up, swigged back my coffee and gathered my books and notes. A ballyhoo of cherry blossom leapt about the wall of Darwin College Fellows' Garden. Dead-looking trees creaked out of the sodden grass, sprouted buds and crackled quickly into the sky.

I'd been working on a cartoon about the origin of numbers.

In the late 1970s a young French woman called Denise Schmandt-Besserat made an astonishing discovery. Forgotten in the storeroom of the Fogg Museum of art in Harvard

was a prehistoric clay purse

from the ancient city of Nuzi,

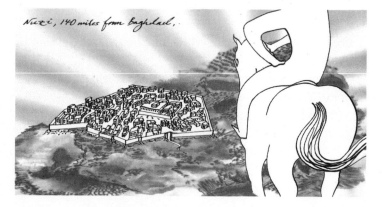

in the country now called

Iraq.

"Uuuugh, aah, errr . . . oh dear!" Simon blustered. "What is the point of this? I don't understand pictures."

"It's the origin of your subject. The purse had an inscription on it that said it was the property of Ziqarru, a shepherd, and contained forty-nine 'counters representing small cattle.' Not that that impressed the Harvard excavators any more than it does you. They broke the seal, found the forty-nine clay pebbles inside, as promised—and lost them."

"Oh dear."

"Exactly. But this French scholar realized that Ziqarru's egg-shaped purse was a simple accounting device, from the dawn of writing. People had discovered other egg-shaped purses containing counters before, but none with symbols on the outside like this. It was the earliest known attempt to symbolize the contents of the purse with abstract marks. According to her, it was the need, by palace accountants, to keep track of animal numbers that led to the invention of writing *and* mathematics. If someone who understood the new marks thought Ziqarru had been stealing animals, all they had to do was check the writing on the outside. And if Ziqarru suspected that person of using the newfangled cuneiform to cheat him, he could break open the purse and prove he was innocent by counting the flock off against the pebbles inside. Lo! Symbolic writing had begun. Next thing you know, it's algebra, calculus and Shakespeare. Writing comes from mathematics, in short, and it all comes from accountancy."

"Oh DEAR!"

"Why 'Oh dear!' this time?"

"No reason," Simon said, sighed morosely and unbent his elbow.

The handles of his duffel rippled down the forearm of his puffa jacket and the bag dropped to the pavement.

"Excuse me!" he gnashed. "I'd like to sit down. Can you remove all this paper?" As he hit the seat he jolted into a better temper.

Simon's most famous ancestor was the Prophet Abraham, of Ur of the Chaldees. Then came Joseph, of the Coat of Many Colors. His son was Manasseh, first mentioned in Genesis, who led one of the twelve tribes of Israel. Next follows 3,000 years of forgetfulness before the family pops back into life on a rolled-up poster in the back room of Simon's Excavation—two shelves along and one up from the television-that-might-have-broken-twenty-years-ago-but-possibly-it's-only-a-fuse:

ASLAN MANASSEH
b. Bombay 1884
m
KITTY MEYER
b. Calcutta 1891

The Manassehs are the leading family of the oldest settled community of people in recorded history: the Iraqi Jews of Babylon, 150 miles from Nuzi and Ziqarru's purse.

Congratulating myself on my willingness to be at the coalface of biographical reportage and, at a moment's notice, drop everything and go to Woking, I walked with Simon from the café, across the park. "The Martian" turned out to be a statue in honor of H. G. Wells' *The War of the Worlds*. According to a tourist leaflet Simon eventually discovered in his coat pocket, it's seven meters high. It looks like a beetle trying to curtsy with its legs stuck in vacuum tubes. There's also a Woking Spaceship embedded in the pavement nearby, and Woking Bacteria, made out of splodges of colored concrete brick.

Battling a wallet from his trouser pocket, in the center of Cambridge Simon boarded a bus to the railway station. He muttered out coins into the driver's cash tray, seized the ticket, held it to the light to investigate it with narrowed eyes, then made for a free space at the back of the bus, bouncing his duffel from ear to ear of the seated passengers. Blank-eyed, belly exposed, his ski jacket rucked halfway up to his chest, Simon threw himself at the seat with a self-congratulatory sigh and let his face settle around his grin.

The face of the woman in front was soured by watchfulness. Simon, though sexless as a nematode, is the fantasy image of a kiddy-fiddler, and this Bruiser Mum had spent her morning proudly dressing up her six-year-old daughter in lash-thickening mascara, gloss lipstick, Primark miniskirt and pink heels.

Standing beside Simon, I took out a notebook, and consulted a list of urgent biographical questions. It is important, with Simon, to select not just the correct wording for a query—one that doesn't contain any banned nouns or adjectives, or lead to outbursts of correction because of a tiny factual error—but also the right context. *PHILOSOPHICAL* questions are best on a Tuesday night. This is because he has returned from his weekend jaunt to Scarborough, via Glasgow, the Isle of Man and Pratt's Bottom, finished his week's backlog of 347 emails: he is feeling expansive and post-prandial. Questions requiring *REMINISCENCE* can be extended as far as Thursday, or broached on country walks through Iron Age hill forts—there is nothing quite like 2,000-year-old battlements, where the clash of Roman legion against shrieking Celt still trembles in the air, to get Simon going on the subject of Ashdown, his junior school.

Bus trips to the train station are strictly for the exchange of *FACTS*.

The scholar of Simon Norton Studies must proceed with delicacy.

"I wanted to ask about your grandfather, Aslan," I began. "He was a businessman, wasn't he?"

"If you say so."

"What did he sell?"

"I don't know."

"According to your brothers it was textiles, but what . . ."

"Yesterday I was in Blickling." Simon pinched his fingers into his wallet and extracted a worm of paper. "Here's the ticket."

"Simon, your *grandfather*. Was it jute?"

He waggled the ticket higher in the air, closer to my face. Four inches long, it had arrowhead shapes cut out at either end, and purple 1970s techno-writing along the length repeating with great mechanical urgency, top and bottom, that it was 1:23 p.m. in King's Lynn, and that Sheldrake Travel was "very happy to have you aboard."

("I do not think that could have been the ticket I showed you. There is no direct bus from King's Lynn to Blickling. But if you prefer to get things wrong deliberately, you belong on the team of a trash publication like the *National Enquirer*.")

"Is there anything special about it?" I asked, too self-conscious to hold the snippet of paper up to the light and try out his squinting trick, without at least some guarantee of reward.

Simon considered for a moment, then shook his head contentedly. "No."

"I was in Blickling last week too," said a fellow sitting beside Simon. The man was resting his chin on his hands, which were in turn piled on the handle of his walking cane; he bounced his head gently. "Lovely hall, and, aaah, the lake. I got there very early and the mist, it was . . ."

"Did you go on any buses?" Simon blurted.

"To the hall," agreed the man, nodding some more, rather slowly, as if tapping the sharpness out of the interruption.

"From?" shot Simon.

"Norwich, I believe it . . ."

"The number X5," Simon declared, and directed a smile of triumph around the bus.

The elderly man was not to be put off: he was a trouper for the cause of discursive memoir. "I think my favorite—I mean, lakes are always lovely, but lakes are lakes, I always say—my favorite was the Chinese Room. Did you see that? That flock wallpaper, it was flock, wasn't it, and that pagoda in the glass cage . . . ?"

"Any other buses?"

"Well, after lunch, we went to Cromer, and had the most delicious brown crab . . ."

"The X5 again. Unless you went on a Sunday?"

"No, let's see, Tuesday, that's it, because then at Wells-Next-the-Sea, the sunlight on the water was sparkling in just . . ."

"Uggh, ah . . ." Simon pulled out a dog-eared timetable from his bag and searched the pages. "Let's see, aah . . . the 73." Spotting that Nodding Man still had a bit of life in him, Simon brought in the heavy artillery, lifted out a second book, which seemed to be compressed from the scrag ends of newspaper, ran his fingers down the index and began darting back and forth between two sections at once. "But you could have taken the 645 and changed at . . . let's see, aaah . . . or, uuugh, aaaghhh, if you'd wanted to go on the steam railway . . .

("Alex! What are you saying? Number 73? Number 645? A steam train? I am sure you have invented these references also. I could *not* have said them. Do you want me to be seen as an ignoramus on public transport?")

". . . which calls at hnnnn . . . King's Lynn, and . . ."

It began to rain. First, a barely visible drizzle, picked out only against certain backgrounds—the black reflections in the windows of the Cambridge Hotel; a middle-distance blurriness when the bus stopped at the crossroads by the Catholic church, and we had a view up to the park. But it might have been nothing more than stripes of movement left in my eyes by the Clint Eastwood action smack-'em-blast-'em-ride-off-into-them-thar-cactus-lands flick I'd watched last night. Next, streaks of water on the window. Finally, drops pounding the metal sill by Simon's elbow in buttercup explosions.

"Getting back to your grandfather, Aslan . . ."

The driver slammed the brakes and swerved to avoid a line of Japanese girls who'd abruptly pedaled across the road in front. The bus was filled with sudden pushes and violent attempts to avoid falling over. I crashed forward down the aisle and fell sideways onto the six-year-old nympho.

"Oi, watch where you're fucking going," growled Bruiser Mum.

Simon, who spends much of his time smiling, smiled wider. He burrowed into his bag and, after much rustling and what

82

looked like punches delivered at the fabric from the inside, re-emerged holding a carton of passionfruit juice, which he upended over his mouth.

At the end of the nineteenth century there were 50,000 Jews—a quarter of the city's population—living peaceably alongside Arabs in Baghdad. Today, according to the latest Web report, there are four—*four* in the entire city. The pro-Hitler Iraqi government expelled and murdered them in pogroms before and during the Second World War. In the late 1940s underground movements smuggled them to safety at the rate of 1,000 a month. In 1951, Israel airlifted 60,000 more from the whole of Iraq and, with the perversity of the self-justified, bombed the rest to try to persuade them to follow. There are today more ostriches in Baghdad than there are Jews.

On one edge of the genealogical poster I'd excavated in the basement is a dedicatory note about Simon's family:

> All probabilities and evidence go to suggest that this community is descended from the ancient Jewish communities settled in Mesopotamia since the days of the Babylonian Captivity, 2,600 years ago ... The purpose, in compiling the genealogical table, is to preserve, in some way, a record of a section of this community.

The very same day that Israel finally declared independence as a refuge for the most persecuted race on earth, Syria, Lebanon, Egypt, Jordan and Iraq launched a combined attack, which the Secretary of the Arab League declared on Cairo radio was "a war of extermination, and a momentous massacre which will be spoken of like the Mongolian massacres and the Crusades."

"Murder the Jews! Murder them all!" shrieked the leading Islamic scholar of Jerusalem.

Sixty years later, a man in London offered a million pounds to any breeding Iraqi Jewish couple who would go out to Baghdad to repopulate the city. "I have a friend who's interested," I enthused to Simon. "What do you think? Her name's Samantha."

"I dislike the name Samantha, so anyone with that name would be unlikely to attract me. Maybe it's because it makes me think of Samantha Fox, the pornography star . . . I may say, I do have a relation with a Samantha. She deals with my tax affairs."

Simon settled into a dead-eyed stare, gave himself a hug with his elbows and went back to looking out of the window: a quiet, euphoric gesture. Until we were on the train, he could devote his entire attention to ignoring me.

Higher up Simon's genealogical poster, closer to the rustle of the Old Testament, the children are nameless, lives are replaced by question marks, but deaths are biblical: a sister to Habebah, "drowned in the Euphrates"; Sassoon Aslan, "buried in Basra"; Minahem Aslan, "childless, in Jerusalem." Before that, Simon's family disappears off the top of the page into the Mesopotamian sand dunes.

At the train station, Simon jolted off the bus to the fast ticket machine in the concourse and pressed screen after screen of glowing virtual buttons. Once he'd finally amassed all our possible discounts, off-peak fares and unexpected mid-journey changes to thwart the local train operators' pricing structures, he stared for a minute at the screen, which was demanding to know how many passengers apart from himself were taking the trip.

"0" pressed Simon, and looked up at me without crossness or dismissal.

* * *

Together, Aslan and Kitty Manasseh had five children, spaced every two years: Maurice, whose wife sneaked off one day when he was out and had herself sterilized; Nina, an old maid; Lilian, who ended up "in Blanchard's antiques shop" . . .

("Do you mean she was for sale, Simon?"

"No! Of course not, *he, he he.*")

. . . in Winchester, childless; Helene, Simon's mother (Gaia among women in that barren setting, because she had three boys); and Violet, a war widow, who added another boy. This man, Simon's first cousin, goes by the name of David Battleaxe.

"You mean he was christened that?" I perked up.

"Not christened, although we do celebrate Christmas. He's Jewish. We're all Jewish," replied Simon. We were on the train now, hurrying down the aisle.

"David *Battleaxe* . . . ?"

"After a racehorse."

"A *racehorse*?" I puffed.

"In Calcutta."

"In Cal*cutta*?"

"One of my grandfather's," said Simon, then lunged left and landed with a thump in a window seat, his bag arriving on his lap—*crucccnchch*—a split second after.

"So you do know *something* about your grandfather," I observed, squeezing past two beer cans into the rear-facing seat opposite, next to the toilet. "He kept racehorses and named his grandson after a stallion. Yet when I asked you what your grandfather did just now, you said you didn't know."

"You asked me what he *traded,* and I said I didn't remember."

The train pulled away, clacked across various points until it found the London tracks, and mumbled past the Cityboy apartments with tin-can Juliet balconies.

"I don't think he did trade horses," resumed Simon, as we picked up speed toward the Gog Magog hills. "Therefore I did not feel that it was relevant to provide that as an answer."

A conductor hurried up to us, clicking his puncher, jutting his chin across seat columns, and demanded tickets and railcards.

Simon had his wallet already prepared, bunched in his fist, and offered up his pass and all the other necessary pieces of colored cardboard in a derangement of eagerness. So many, the conductor needed an extra hand to deal with it all: the outward from Cambridge to Wimbledon via Clapham Junction and Willesden Junction covered by one set of reduced-fare permits; a continued discount outward from Wimbledon to Woking, with "appropriate alternative documentation." As the man sifted through these triumphs of cunning, Simon's face was suffused with expectation. The conductor adopted a bored expression and punched whatever suited him with a machine that pinched the paper *hard* and left behind purple bumps. Simon snatched the pile back and studied the undulations with satisfaction.

Another cousin I'd noticed on the family tree was called "Bonewit." This woman appears on the fecund side of the family. It's difficult to count the tiny layers of type on that half of the poster: seven children to Joseph and Regina; eight to Isaac Shellim and Ammam; ten—no, twelve—wait, my finger's too fat for the tiny letters, *eleven*—to Shima and Manasseh: Aaron, Hababah, Ezekiel, Benjamin, David, Hannah, Esther . . . a rat-a-tat from the Pentateuch. Fifteen kids! to Sarah and Moses David. By the time they got to Gretha Bonewit, their seed was worn out.

"Bonewit?" said Simon, interrupting. "'Wit' is Dutch for 'white.' I've got a Dutch dictionary in here."

As the train passed Addenbrooke's Hospital, Simon's attention swerved, to gout. Jolting his hand out of the foreign-dictionary sector of his duffel, he sank it back in six inches farther along and two inches to the right, and extracted a scrunched-up Tesco bag containing tablets. Allopurinol, for gout; Voltarol, for swelling (though it's bad for his kidneys); Atenolol for blood thinning. He washed a selection down with more passionfruit juice and returned to dictionary hunting.

"Simon, *why* have you got a Dutch dictionary?"

"Why shouldn't I have a Dutch dictionary?"

"Do you have a Mongolian dictionary?"

"No."

"Do you have a dictionary for roast chickens?"

"No."

"Well, then, why a Dutch one?"

Simon's mother, grandfather Aslan (far right) and Battleaxe.

"Because," he honked, triumphant that the answer had got such assiduous courting, "I . . ." But at this point he found the book in question. "Let's see, aaaah, hnnnn, bonewit, bone, bon . . . ooh . . ."—his eyes lit up—". . . it means 'ticket voucher.'"

Simon will rot his floorboards with bathwater, immure his kitchen surfaces in Mr. Patak's mixed pickle and hack his hair off with a kitchen knife, but he is never unkind to maps. Returning the dictionary, Simon burrowed a foot and a half to the left and cosseted out an Ordnance Survey "Landranger." He shook it into a sail-sized billow of paper, then pressed it gently into manageable shape.

Outside, the rain was frenzied. It clattered against the roof and ran in urgent, buffeted streaks along the glass. The flat lands of Cambridgeshire swelled up into a wave of hills.

When I looked back at Simon, a banana had appeared in his hand.

"Right, your granny. Why did she live in Woking but your grandfather stayed in Calcutta?"

"I have no idea." Simon looked up from his map and considered the point. "Isn't that the sort of thing married couples do?"

"Was there a huge argument?"

"No, oh dear, I don't know."

"Did he have a harem?"

"Huuunh. Should he have?"

Ordinarily, I like to record all interviews, because it's not just the words that count but the hesitations and silences. But this opportunity had occurred without notice, and I didn't have my voice recorder.

I decided "Hnnn," "Uuugh" and "Aaah" should be noted as "H〰," "U〰" and "A〰." Stage directions "pained," "dead-eyed" and "yawning" to be added as appropriate.

"OK. How about this: why did your ancestors leave Iraq for Calcutta in the first place?"

"Oh, dear, no. No, no," Simon replied. "I can't possibly remember that. A 〜〜. How can I be expected to remember what happened before my birth?"

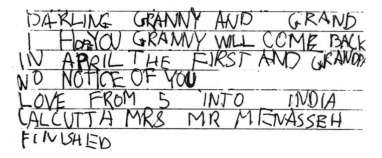

Letter to grandparents, from Simon
(signing himself by number 5) aged 5.

In all of Simon's recollections Kitty hobbles. After emigrating in the year he-doesn't-know-when, leaving behind he-doesn't-know-why her husband, Aslan, she bought a he-doesn't-know-what-type-of-house in Woking with a bamboo plantation.

"Bam*boo*?"

Simon doesn't-know-how—I mean, doesn't know how—it got there. Every day, until her nineties, she dragged herself round, at first flicking gravel off the petunia bed with her walking stick; then, in her final stages of life, pruning the box-hedge parterre from her wheelchair, pushed by a daughter or a friendly guest.

"One of her legs was broken," is Simon's explanation for the hobble.

"Permanently?" I asked, and paused. "Which one?"

Simon thought carefully. "H 〰️ (pained), the left." Then he considered the problem a moment longer: "A 〰️ (aggravated), the right."

Another bout of concentration.

"They alternated. Would you like some Bombay mix?"

626 MOR·R$SB E S
3 12 1957 V W,

MY DEAR MUMMY WHEN I HAD MY
HAIRCUT I THOUGHT THE HAN AT THE
HAIRDRESSERS WILL CUT OFF MY HEAD AND
THE WOULD I BE DEAD?
AND WHEN IT IS SATURDAY I AM SO
VERY PLEASED AT WOKING BECAUSE I
STAY NINETEEN HOURS A DAY AND
SUNDAY WHEN I DO NOT GO TO
WOKING I AM VERY AND .
NEVER STOP SAYING VERY.
CROSS WITH YOU LOVE
5

Letter to his mother, age 5.

When she wasn't in the garden, Kitty sat in the front room overlooking the croquet lawn and played bridge. Her entire last thirty years seem to have been wasted on hobbling and cards.

Grandfather Aslan was "fairy-like." Once every few years he appeared in London for a week, then disappeared. "Feeew-ff, just like that." The rest of the time he remained in Calcutta, the very successful dealer in . . . Simon still-doesn't-know-what.

That's it. There's no point in prolonging this ancestral agony.

*11

Introducing

Just as a square can be rotated through four turns to get it back to where it was to begin with, and the results laid out in a Group Table, the same approach can be applied to every regular shape. The size of the table you need to draw depends on how many operations have to be performed before you've exhausted all the possibilities and ended up back where you started. An equal-sided triangle can be manhandled three times before it's back on its feet:

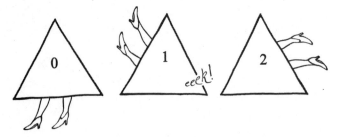

As before, these represent the *act* of turning Triangle. The trick of the game is to find all the ways you can fiddle with Triangle and yet leave it looking just the same afterward as it did before you began:

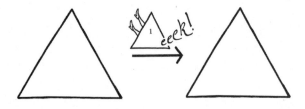

And (again as before, with Square) these turns combine in the most obvious way . . .

In words, turn Triangle once, then turn it again, and the result is two turns: one plus one equals two. It is easy to spin Triangle head over high-heels, if that's what you want:

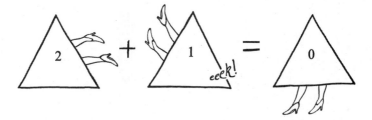

$$2 + 1 = 0$$

(two turns, followed by one turn, returns Triangle to its original position)

Remember, in Group Theory, turning a regular shape right round is taken to be the same as doing nothing at all. Full, completed turns don't get totted up. It's only the overall adjustment that matters:

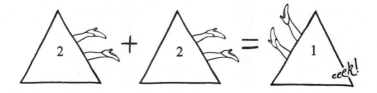

$$2 + 2 = 1$$

The corresponding table (which, as with Square, looks like a pint-sized sudoku table) is therefore:

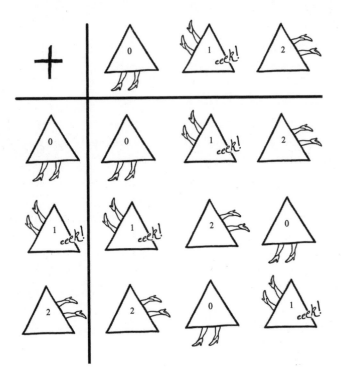

Once again, we've got through a mathematical section with a suspicious lack of awfulness, like someone who's committed a crime in the woods.

Is that all there is to it? Was that really mathematics?

The mist in these woods is hushed. A distant leaf clatters among the branches like a falling pin.

Let it be whispered: a saucy chapter is approaching.

12

I had a camera once. When I found it wasn't working,
I had a sigh of relief.

Simon

The name "Norton" is a fake. Simon's paternal grandparents,
from Germany, died before he was born—"No Nazis involved."
His paternal grandfather anglicized the name from Neuhofer—
i.e., New Towner, usually turned into Newton—then fiddled
the result to sound less desperate. The only thing we can say
for certain about these people, Simon pronounces
sententiously, "is that when the surname was being used in
Germany, the one place the Neuhofers did *not* live was in any
village called Neuhof." Then he sits back and stares at me
happily, waiting for understanding to dawn.

The fact is, for all the interest Simon takes in his ancestors,
you'd think he was bred in a flowerpot.

The rest of that day's gentle journey through the Southeast
of England was taken up with Simon tracing our route on the
Ordnance Survey map with his forefinger. He is extraordinarily
bad at this. Invariably, he knows where we were ten minutes
ago, or where we are about to be after the next level crossing;
it's where we are now that boggles him. He's in a constant fret
to find the spot. Every fifteen or twenty minutes he'll succeed,
look up, catch my attention and desperately point through the
window at a "site of special historical interest" several seconds
after it's disappeared from view. Simon's general attitude to the
third dimension is wary and defiant. It is always playing tricks
on him. Two and 196,883 dimensions are the ones Simon

prefers. He can spend the afternoon in his bath with twenty-four dimensions and suffer no ill effects at all. But every time he attempts to locate a three-dimensional racecourse or a well-known housing estate that had once appeared in the *New Scientist* because it was built in the shape of a snowflake, a tall bush has got in the way, or the train is making an unexpected detour around a hill, or everything is pitch black because he's actually in a tunnel.

Between these annoyances, Simon devised games. "What is the minimum number of London Underground stations you have to go through to get all the letters of the alphabet?"

This is an old one of his.

"Now, let me see, H 〰〰," I suggested, taking out my notebook and using appropriate Simon noises to help me along. "You take the Circle Line west from King's Cross, which gets you, A 〰〰, let's see now, Baker Street, that's a, b, c, e, g, i, k, n, o, r, s, t. Then take the Bakerloo Line to . . ."

Simon doesn't use his fingers or paper when he's playing this type of memory trick. With exactly the same expression on his face that he has when analyzing calculations, he parades stations through his recollection, snipping off letters. When my attempt crashed into the buffers after the third station, he began again at King's Cross: "Northern Line to . . . then change at . . . next the Jubilee Line to . . . U 〰〰 (yawning) . . . St. John's Wood—fourteen stops in all."

The point about dimensions holds here too. The Tube is one of the few places Simon never gets disoriented, because under London there are no landmarks to miss. Down there, the third dimension is made up entirely of shiny colored tiles and adverts for Jack Daniel's.

Another game was "What's the smallest portion of the map that's got every letter of the alphabet in it? It was a visit to Cornwall that made me think of this one," he said with pride.

"There are lots of places with a 'z' in them down there. Ohhh, look, there! THERE!" Simon began stabbing at the window of our rail car. Flashed past on the other side: a break in the hedgerow. Beyond it, tarmac, the glint of metal, a slash of something red and fat. Then it was gone. Simon beamed with satisfaction. "Weybridge Station—there used to be a Bridge Club there," he pronounced.

It's Simon's belief that all children should have compulsory lessons in orienteering and how to use public transport, "so they can learn to enjoy the countryside before they reach driving age, and as a result never want to learn to drive."

There is another advantage to this plan for enforced education: the idea of representing something in different formats is a fundamental one in mathematics, so understanding how to read maps and bus timetables would also help children appreciate the concepts behind their algebra homework.

The only other ancestral snippet I extracted was that every week of the year, on Friday, and also at Christmas, Simon's father drove the family down to Woking in the Bentley to visit Kitty Hobbler. It is the one point about those days that remains distinct in Simon's mind—not because of the droning regularity of the trips but because, strapped in the backseat in a swirl of petrol fumes and his mother's "particularly pungent" French cigarettes, "that's when my dislike of cars began." Cars are "smelly, they kill children, they destroy the planet."

"So do buses. Why pick on cars?"

"Cars are worse. Cars corrode mankind. Incidentally, have I told you my method of remembering the names of the lanthanides?" The lanthanides are soft metals that burn in air, and appear as a row of chemical elements dangling off the bottom of the periodic table.

"Loathsome Cars Produce Noxious Polluting Smelly Exhaust Gases: They Destroy Human Environments, Take Young Lives."

And for the actinides, another dangling row of metallic elements, this time radioactive:

"Avoid These Perils. Use No Private Automobiles. Cars Bring Complete Enslavement For Mankind, Not Liberation!"

Simon, aged two.

"Now, about your grandparents . . ." I began, determined to return to biography.

"Have you seen this?" Simon burst out.

He dragged out a Philip's atlas: ring-bound, supersize. The two most important books in Simon's life are both atlases: the *Atlas of Finite Groups,* which made him famous throughout the world of Group Theory, and this grubby guide to the roads of Britain. Opening the cover, he flicked past a couple of pages. The paper, through his fingering, has acquired a dirty down.

If his house were burning down, this Philip's road atlas, says Simon, is the only thing he would run through the falling timber to retrieve.

Each time Simon travels along a new route on a bus, he traces over the relevant line in this book with a pencil. The Midlands, Kent and the Lake District are coated in smudge. Scotland, with the exception of a few contrails of graphite around Edinburgh and Scrabster, is largely crisp and clear. Along inland borders, his squiggles tend to get lost because on Philip's maps the county divisions are gray. The pages around Cambridge, submerged in carbon, have torn loose and flop about the middle of the book.

Simon is in the process of transferring the shading from one Philip's atlas to a new one. It isn't easy to discover why this record of the-tarmac-I-have-known he's covered matters so much to him.

"Is it a sort of stamp album? Are you hoping to collect all the roads of Britain in it?"

"No."

"A record of every bus journey you've ever made?"

"No."

"Would you like it to be like that? Would that be the ultimate? Is it your memory? Are you being peculiar and trying to write out a sentence in bus journeys across the surface of the country?"

"No!"

One of the things I enjoy about Simon is his revulsion at my attempts to make him novelistically tidy.

Some of the pencil lines in his road atlas shoot into the sea. An amphibious bus service, was it? An inch or two free of land they start to arc, as though compensating for the curvature of the earth, and become obvious boat trips, ending up at the Isle of Man, Skye, Eigg, Wight.

Within these islands, none of the roads are shaded.

Why else would Simon have gone to those isolated spots, except to cover all roads?

"Have these islands got a mental tick beside them?"

"No."

The sticker on the cover of the atlas said £6.99. "Actually, it was £1.99 in Budgens," he confessed, folding the book shut, delighted at the low cost of pricelessness. Then he picked up his map again, darted glances through the windows on either side of the rail car to remember his point of departure from three dimensions, and slid back into silence.

Fretful with impatience, Simon bustled down Woking platform as fast as his duffel and gout would permit, noted with satisfaction that the local bus service was three minutes late and, worried that those minutes might suddenly pick on us instead and delay all his plans, made quickly for the center of town with me trotting behind.

The Martian is at the end of a dreary pedestrian walkway, its legs buckled with despair at finding that it's traveled sixty million miles, the last hope of a dying civilization, and ended up in Woking. In *Woking,* being (as the photo shows) punched by a lamppost and chased by fake Victorian bollards.

Standing on a colored microbe, Simon admired the sculpture, consulted a street map and ate a pack of Bombay mix he'd found wandering about his bag.

"Do you like this Martian?" I asked.

"It's all right," he said.

"Do you like Woking?"

"H 〰. It's all right. When I got older, I used my grandmother as my springboard. I used her to bounce all over the Southwest."

We walked up the hill, and through a brick gateway into a private compound of 1930s houses, set, as in the American Midwest, among redwoods. The private roads rested on grass, between daffodils; the buildings, three times the size of city houses, were stockaded by laurel hedgerows and topiary yews.

"H 〰〰 (with contentment) this is it, my grandmother's house," announced Simon.

But the moment we stepped in to look around, Simon got frightened that we were trespassing, and ran away.

*13

Saucy Miss Triangle!

The Outrage so far: An equal-sided triangle is symmetrical. It looks unchanged when you manhandle it through a third of a turn, two-thirds of a turn, or turn it all the way round.

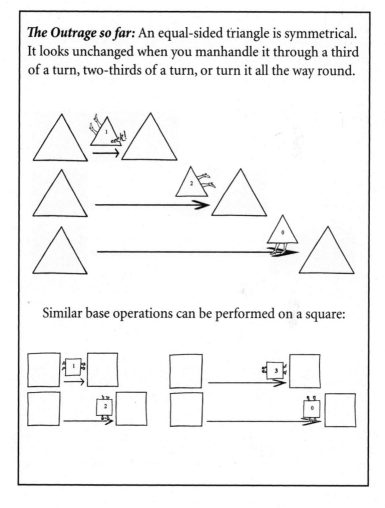

Similar base operations can be performed on a square:

If you have the energy to perform two or more of these misdeeds at once, you can keep track of the overall effect with the help of a grid that looks like a sudoku table:

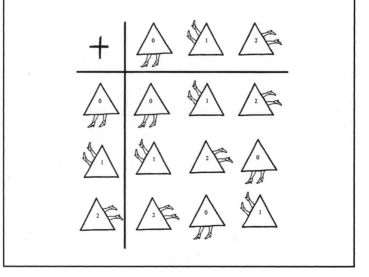

There's another symmetry operation—nothing to do with rotations—that allows us to have our way with Triangle, and still leave it to all appearances untouched. We can flip it back to front:

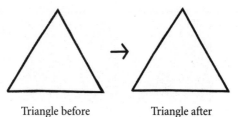

Triangle before   Triangle after

Inwardly, Triangle may adjust its spectacles and primly pat its skirt, but outwardly it looks unperturbed. We record the act in our diaries by a new symbol:

Remember: this is not the triangle itself. This is what we have done to it:

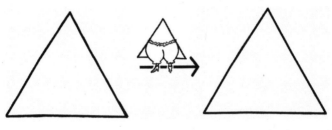

Triangle before   Triangle after

The embloomered shape represents the *act* of turning an equal-sided triangle over:

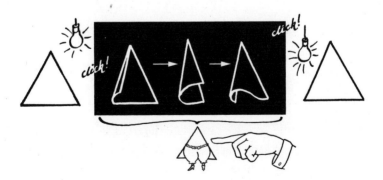

Irregular triangles, naturally, are not so circumspect. Flip one of these misshapes over and it shows very definite signs of adjustment. Anyone watching immediately spots what you're up to. There's none of Triangle's finishing-school ability to keep the secret under wraps:

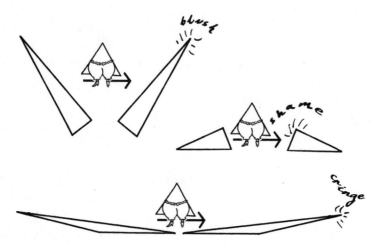

If we add all the ways we can flip an equal-sided triangle—

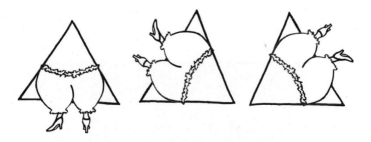

—to the three ways we can rotate the triangle, and construct a Group Table to keep track of how all these symmetry operations interact, we get the following lumbering object:

Next, a page for ladies:

We can do a similar flipping operation with Square:

Turn off the light, flip Square over, turn the light on again, and Square looks delightfully unchanged.

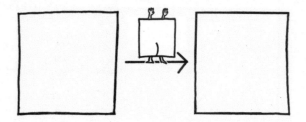

Now we're almost there. The critical, central idea—Subgroups—discovered by a French schoolboy in the nineteenth century. From this child's revolutionary insight spring most of the complexities of Group Theory and a first glimpse of the Monster.

14

I don't like your books, Alex. I don't like your mother's books either.

Simon, to the author

Humming odiously, Simon expressed satisfaction.

Hnnnnneeeuuucch, hmmm, tllahdlllah, hnnnnn.

Outside Woking, on the way back to Cambridge, we'd stopped off at a National Trust country house to listen to a concert of Beethoven sonatas. A Japanese pianist—scraped-back hair, keyboard ribs showing through the V of her dress—stepped up to the piano, bowed, sat down, stood up, smiled at the audience, smiled at the parquet floor, smiled at the portraits on the walls, sat down a third time . . .

The piece was the Kurfürsten Sonata in E flat major, WoO 47.

—ambled her hands into mid-air—

"The first of his posthumous sonatas," said Simon in an attempt at a whisper.

. . . and hit the keys in a sprightly fashion. As you might expect from the first attempt by a corpse, the Kurfürsten is a simple piece (written, in fact, when Beethoven was thirteen, but added to the catalogue of his works only after his death). It sounds like a mouse let loose in a toy room.

Fidget maestro at other times, Simon was clamped in obedient stillness. Until cramped. Then he twisted and arched through the final bars, his eyes fixed, colorless, dead to the world, looking—*hnnn!*—inside himself—*uuhhgh!*—at his ache.

Simon has an upright piano in the back room of the Excavation, keyboard lid open and Beethoven's Moonlight Sonata on the stand. The last time those keys made a noise of any kind was eight years ago, when Simon accidentally knocked the chessboard off the top of the piano and the avalanche of pieces cudgeled a few notes on the way down.

Beethoven is Simon's favorite composer. He enjoys the "unexpected modulation" and the fact that "as with impressionism, the melody does not include every note." His cassette tapes, stained heavily with brown splotches, are arranged in two rows, neatly: Beethoven Sonatas Nos. 14, 22 and 29, played by Sviatoslav Richter; Beethoven Klaviersonaten, Nos. 11, 21, 27 (Buchbinder); Beethoven Sonatas, Nos. 28 and 29 (Maria Yudina); Beethoven, Violin Concerto in D (Milstein); Beethoven, Symphony No. 3 in E flat; Beethoven, Piano Sonatas Ops. 2, 49, 81a (Alfred Brendel); Beethoven, Symphonies 1 and 6 (Norrington); Beethoven, Piano Sonatas, Moonlight, Appassionata, Pathétique. Eight Gilbert and Sullivans; two Flanders and Swanns; and one (which I still haven't got round to investigating) called "Mud in Your Eye"!

("I didn't know I had that," says Simon. "It's about canals.")

Simon doesn't listen to music at home. He doesn't listen to the radio at home. Home is for silence.

During the interval, Simon fell into conversation with a retired piano teacher. Bad pianists play "along a line," she said. They don't stray out into nuance, interpretation and expression. They get trapped, like tired artists and tired biographers and tired mathematicians, in a rut. It seemed to her that one of the three pieces this afternoon had been played a little "along a line." Doesn't Simon agree?

But Simon never likes to be rude about anyone. Or to make *requested* aesthetic judgments. So he hurriedly descended into squirms. He had heard the Kurfürsten, he confessed, "played faster," then he edged toward the refreshment table.

"When I was at a conference in Santa Cruz," he popped back a minute later, "I asked a German there if Beethoven's 'Moonlight Sonata' could equally well be translated as the 'Moonshine Sonata,' and he said that it could."

"My girlfriend," I observed, "saw an art installation in Chichester, at which they beamed a recording of the Moonlight Sonata at the moon, let it bounce off the surface, then replayed the noise that was reflected back to earth after it had been picked up by a satellite dish. They said it was the Moonlight Sonata, as performed by the moon. She said it sounded jolly peculiar. All the bumps and lumps of the moon had distorted it. I don't think, Simon, you're meant to take five chocolate biscuits."

Just because Simon compares his favorite composer to impressionism, it doesn't follow that he likes impressionism. He doesn't. He doesn't like any art, with the exception of "some modern stuff," which turns out (predictably, for a mathematician) to mean M. C. Escher. On his jaunts he occasionally inserts himself into a gallery and scurries around the walls, nodding off whatever paintings he passes, "Because one ought to," but "I don't enjoy it." In particular, he doesn't like portraits or cartoons.

His own early artworks, produced when he was ten years old, are startling: exuberant, abstract, artistically poised, made from lurid combinations of tissue paper. I discovered them in the Excavation squeezed between the groyne corner cupboard and the chest of drawers with the returnable Tango bottles. What makes them especially interesting is that the shapes he uses for these compositions are the shapes we've been looking at in the mathematical sections of this book: the founding forms of Group Theory.

Untitled #1, Yellow, red and brown squares.
Mount: evening-blue sugar paper.

Untitled #2, Multicolored squares and circles and
sticky shapes. Mount: shocking-orange sugar paper.

The emphases on the clam-like, suggestive hiddenness of patterns (*Untitled #1*); interdependence (*Untitled #2*); and balance, not simple regularity (*Untitled #3*, below), capture the essence of Simon's eerie mathematical insights fifteen years later.

Untitled #3, Yellow (circle), pea green
(square), on rectangular field of blue shades.
Mount: shocking-orange sugar paper.

After the Beethoven concert ended, the retired piano teacher drove us to a village train station. Kind gestures such as this are always falling in Simon's way, but he never knows what to do at the end of them.

"U 〰〰," he grunted when we got out of the woman's car.

"A 〰️ ." His eyes tightened, his shoulders stiff with impending duty.

"H 〰️ ." He backed up—tried to remember what the duty was.

"Thank you," he recollected at last, but only because I'd said it a second before. "Goodbye," he added, overhearing me again. "As I say . . . A 〰️ . . . can we go now?"

He bolted off to the platform.

"There! That's where I first picked up . . ."

I glanced up from my seat in the train and caught the last glimpse of a platform with a picket fence and hanging baskets rattling past in the evening light.

". . . the magazine *Thirdrail*. Published by British Rail Southern Region, to encourage rail travel." It was a tough objective. The issue Simon discovered in the ticket hall as a little boy was Number 6, the last produced. After that, the editors gave up the battle against the rise of car travel: the magazine folded.

"Here!"

A second small station: two guards were watering a splosh of tree in a cask. "Merry Makers Mystery Tours!" shouted Simon, as the train went in a tunnel and the batter of rails was replaced by an oceanic rush of annoyed air. "My first leaflet about them: that's where I picked it up. Kept me occupied during the 1970s."

For twenty miles around his grandmother's house in Woking the countryside is marked for Simon by autobiographical signposts: asexual, calm and related to paper.

"Incidentally, if you *do* have to use the derogatory word 'excavation' to describe my basement, you could emphasize the sense in which it is very apt, namely, that you have excavated a lot of information about me there. Then you could feed in something about my interest in archaeology, and maybe you could mention Woking Palace, in view of the importance of Woking in my life—saying that, during all the decades during which I visited Woking, I never heard of it!"

At Kingston upon Thames, he fell asleep.

Power cables, veering and swooping. Over the points they hit jumbles, great knottings, new cables crashing in from all sides, frenzied dogfights—and out flashed two or three solitary lines—free! unencumbered!—and against the sunset the joyous lunges alongside our train car began again.

I felt a biographer's contentment. What had we achieved today? Seen a statue; listened to sonatas; run away from a house. Not much. But something in these small events seemed to characterize Simon and his by-and-large unremarkable, happy life. I didn't want to think closely about what this something was. Expression might have done it harm.

I took out a book I'd been reading: Havelock Ellis's *A Study of British Genius.* Havelock Ellis formed his conclusions about the origins of genius by selecting 1,030 "eminent men" from *The Dictionary of National Biography* and subjecting their entries to a vigorous attack of statistical analysis, investigating the intellectual influence of parentage, upbringing, schooling, place of birth, number of siblings and place in sequence, illness early in life. With one exception, none of his conclusions applies usefully to Simon. He believed men of genius were often unhappy or ill as children (Simon was not); were averse to ordinary work (which Simon certainly is, but so are burglars and politicians); had one brilliant parent or at least one who was dead (Simon had neither, although his mother was clearly much brighter than his father); and noted Aristotle's insistence that all men of ability are melancholy (Simon is not) and have a mixture of insanity in them (Simon does not).

The genius-insanity cliché is so absurd it's worth quoting Havelock Ellis's dismissal:

"Among poets and men of letters, *of an order below the highest,* insanity has been somewhat apt to occur," he agreed, but not among the truly great, or mathematicians. Antiquarians, he added, possess "marked eccentricity almost amounting to insanity," but these people turn out to be otherwise so low on brainpower "that the question of their inclusion in my list has been a frequent source of embarrassment."

Havelock Ellis's mentor, the brilliant sociologist Cesare Lombroso, thought prodigies like Simon were related to murderers and bank robbers, a morally degenerate form of mankind.

Simon's only mental pathology is an excessive desire to obey local housing law.

If Simon were in charge of anything, it would grind to a halt under his honesty.

Simon says he's a fluke of genetics. Every birth is a gamble by Nature, a throwing in the air of infinite possibilities. The good freaks that result are as random as the bad. In Simon's case, "the molecules settled in my favor. Neither of my brothers is particularly intelligent."

(One of these brothers once tried to convince me he'd scored 120 percent in a school exam.

"But you can't score 120 percent," I'd complained.

"They had to hush it up.")

In "Why productivity fades with age: The crime-genius connection" (2002), Satoshi Kanazawa applied Havelock Ellis's technique to *The Biographical Dictionary of Scientists*—but this article is such garbage! Kanazawa makes a couple of guesses about the high point of each scientist's genius, shows that the ages over which these pinnacles of brilliance occur follow a similar pattern to the ages at which criminals perform their worst deeds, and then—his bizarre conclusion?

That geniuses and gangsters have the same underlying motivation: the urge to seduce women.

Simon is not interested in women.

After 300 pages, covering the fourth to the nineteenth centuries and including the additional reading of 300 further biographies, the only common characteristic of genius that Havelock Ellis could find that is relevant to Simon is that these brainy men all have gout.

Moreover, the eminence of these gouty subjects is as notable as their number. They include Milton, Harvey, Sydenham, Newton, Gibbon, Fielding, Hunter, Johnson, Congreve, the Pitts, J. Wesley, Landor, W. R. Hamilton and C. Darwin, while the Bacons were a gouty family. It would probably be impossible to match the group of gouty men of genius, for varied and pre-eminent intellectual ability, by any combination of non-gouty individuals on our list. It may be added that these gouty men of genius have frequently been eccentric, often very irascible . . .

15

When Noël was still two the doctor pronounced that his brain was much in advance of his body and advised that he should be left very quiet, that all his curls should be cut off and that he was to go to no parties.

Cole Lesley, The Life of Noël Coward *(1976)*

"Two, two, two . . ."

Simon, have you had a stroke?

"Two, two . . ."

We were sitting in the Excavation, Simon eating Mackerel Norton again, but this time the authentic version, with Chinese-flavor packet rice. The stench creosoted your lungs. It's as though Batchelors has thrown the Chinaman in with the rice. I was in the middle of the room, balanced on a dining chair that often drifts on the sea of rubbish, several feet north of the corner clothes cupboard. Simon was rocking on his bed, the stained and crumpled blankets pushed aside, his eyes crimped with delight.

". . . two, two, two . . ."

Simon! What's going on?

He forked up more fillet and boiled Chinaman from the plate on his lap. ". . . two, two, two . . ." exhaled in a cloud of fishy bits.

Somebody shoot him!

". . . two, two, two, two . . ." cried Simon, louder.

Simon's first ever mathematical memory is of sitting on his parents' sofa, working out the value of two to the power of thirty, that is:

$$2 \times 2 \times 2 \times 2 \times 2 \times 2 \times 2 \times 2 \times 2 \times 2 \times 2 \times 2 \times 2 \times 2 \times 2 \times$$
$$2 \times 2 \times 2 \times 2 \times 2 \times 2 \times 2 \times 2 \times 2 \times 2 \times 2 \times 2 \times 2 \times 2 \times 2 = 2^{30}$$

He doesn't know why he started this snake of digits. He remembers not being able to stop. One moment he was fidgeting quietly on the sofa cushions; the next he was soaring into the stratosphere of the thousands, and lo! "My life as a mathematician had begun."

$$2 \times 2 \times 2 \times 2 \times 2 \times 2 \times 2 \times 2 \times 2 \times 2 = 2^{10} = 1{,}024$$
$$2 \times 2 \times 2 \times 2 \times 2 \times 2 \times 2 \times 2 \times 2 \times 2 \times 2 = 2^{11} = 2{,}048$$
$$2 \times 2 \times 2 \times 2 \times 2 \times 2 \times 2 \times 2 \times 2 \times 2 \times 2 \times 2 = 2^{12} = 4{,}096 \ldots$$

He still likes twos. He enjoys how they underpin half the number system, and the rest of mathematics teeters on top.

". . . two, two . . ."

The Monster (as Simon deftly observes) is (take a *deep* breath) "a group of characteristic 2 type, with involution centralizer, structure $2^{1+24}Co_1$, where the group at the bottom of the normal series of the centralizer is of order 2; the group in the middle of order 2^{24}, and if one divides the order of the top group of the series by that of the Conway Group, one gets 2^{25}."

That day, sunk in his mother's sofa, he noticed that 2^{25} (which is to say two, multiplied by itself twenty-five times over) equals 33,554,432, "and I liked the fact that it began with three pairs of digits."

". . . two, two, two . . ." he continues now.

Stop, Simon!

". . . two to the twenty-six, two, two, two . . ." he sped up. "TWO!"

Deflated with satisfaction, he downed three congratulatory mouthfuls and bounced his lips to endure the heat.

"Two to the power of thirty is one, oh, seven, three, seven, four, one, eight, two, four." Above the mid-hundred thousands, Simon rarely calls a number by its full name. He raises his eyes up and slightly to the right, and reads each digit off, as though they floated there, projected out of his mind onto a screen in the sky. "Excuse me, this rice is too hot. Can I use that napkin to blow my nose?"

When Simon first did this multiplication as a child, he liked the patterns the numbers made.

"What do you mean, 'patterns'?"

"Two to the power of twenty-three was my favorite":

$$2 \times 2 \times 2 \times 2 \times 2 \times 2 \times 2 \times 2 \times 2 \times 2 \times 2 \times 2 \times 2 \times 2 \times 2 \times$$
$$2 \times 2 \times 2 \times 2 \times 2 \times 2 \times 2 \times 2 = 2^{23} = 8{,}388{,}608$$

"Why?"

"It spelled 'scissors.'"

Mathematics is the study of patterns. We use the word "two" to describe the pattern of "twoness"—the pattern we see if someone shows us first two of Simon's socks with holes, then two empty mackerel tins, next two bus tickets to an orchid lecture in Pembrokeshire, after that two plastic bags of British Rail "Weekend Rover" holiday brochures bought in Torbay . . . The existence of this "twoness" pattern (and also the patterns for "oneness," "threeness," "forty-sevenness," etc.) gradually led primordial man to develop the idea of a sequence of whole numbers, stretching from the small values, such as 1, 2 and 3, needed to tot up woolly mammoths and caves, through the midland quantities—10, 15, 20—better suited for sheep and birds' eggs, and on to the infinite reaches of God.

What Simon means by saying that he likes the "pattern" that two to the power of twenty-three made "because it spells 'scissors,'" is that the answer:

$$2^{23} = 8,388,608$$

has its digits set out in the same arrangement as the letters in the word "scissors."

Replace the S's with eights:

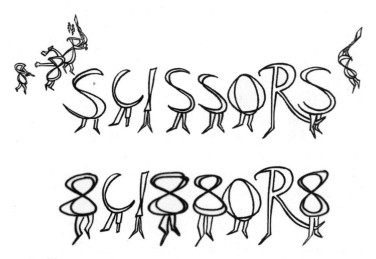

And let 3, 6 and 0 swap with the remaining letters, as appropriate:

"So you see," said Simon, who had meanwhile gone to the kitchen for more supper, "8,388,608 therefore has the same form as—"

"No it doesn't," I interrupted. "It can't spell 'scissors' because it still leaves the 'c' unaccounted for. 8,388,608 has seven digits, and 'scissors' has eight letters."

"It does when you spell 'scissors' the way I did: S-I-S-S-O-R-S."

"But that's cheating! I could make any number I like equal any word if I took that approach. It makes a nonsense of the whole game."

Simon re-emerged from the kitchen in a fresh waft of headless fish and boiled Chinaman.

"It doesn't when you're four."

Another puzzle-book game Simon played as a child—utterly pointless, vital to his mathematical development—was to turn phrases into sums. It's important to understand how these games work. They're the next step up from "sissors," and they confirm (I think) the arrival of something critical—a sort of mischievous joy, a playfulness; I'm not quite sure what to call it—to Simon's genius.

Simon's love of mathematics as a child had nothing to do with logic. It was aesthetic and jocular. He enjoyed the subject in the same way a theatergoer applauds an energetic musical.

The most famous phrase-into-sum was invented in 1924 by Henry Dudeney, a civil servant:

$$\begin{array}{r} S E N D \\ + M O R E \\ \hline M O N E Y \end{array}$$

The aim is to replace each letter with a unique number, and make the sum come out right. After a great deal of shredded paper and pencils hurled at the wall, you discover that the only possible substitution is:

$$\begin{array}{r} S E N D \\ + M O R E \\ \hline M O N E Y \end{array} \longrightarrow \begin{array}{r} 9\,5\,6\,7 \\ + 1\,0\,8\,5 \\ \hline 1\,0\,6\,5\,2 \end{array}$$

These puzzles are like one of Simon's anecdotes. Just when you think you're getting somewhere, they're abruptly over and you're nowhere at all. The intriguing message you started with (why "SEND MORE MONEY"? Is Mr. Dudeney on the run from the police? Is he kidnapped, tied to a chair, with a pistol at his head?) turns abruptly to dust: it's nothing more than 9,567 + 1,085 = 10,652. So what? You're winded of excitement.

Computers are ideally suited to this type of problem. They chew through all the options for the letters at vast speed, and spit out the answer in milliseconds. The good schoolchild puzzler likes to discover shortcuts. He wants to exploit patterns to help him out. The most important thing is not the answer, but reaching the answer with slyness. He wants to creep round the back of the puzzle and give it a sharp pinch; then he'll puff up his chest, pull at his shirt collar and consider himself a clever swell. A good mathematician has a lot in common with a seaside hoodlum.

Immediately, for example, he can tell that the "M" in "MONEY" must equal 1. This exploits a basic pattern in elementary arithmetic. If you take two single-digit numbers—say, 8 and 3,

that add together

to give a *two-digit* number

then that longer number will *always* begin with a 1 (3 + 8 = 11).

Like so many rules in arithmetic, this sounds fastidious and clunky when described at length, but in fact everybody knows it instinctively. It's the rule that makes you furrow your brow when you're being cheated at the coffee shop. "Three quid for a croissant and £8 for a cappuccino makes forty *WHAT*?"

The "M" of "MONEY" in the word puzzle *must* therefore be equal to 1.

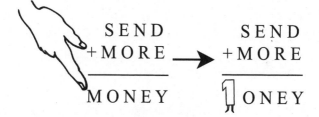

If it didn't equal 1, a pattern of mathematics would be broken, gout pills would walk by themselves, buses fly on wings, headless fish leap up from their tins and chase Simon around the kitchen. And if that "M" must be a "1," then (by the rules of the game) so must the other "M":

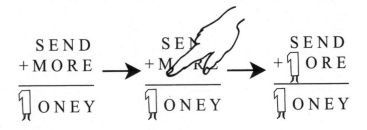

And so on. Pinch by pinch, the mathematician pesters the puzzle until it bursts into tears.

The first puzzle of this type that Simon invented, as a schoolboy, was:

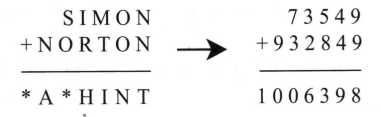

Which is feeble. The words should make up an interesting phrase. Why is Simon Norton a hint? A hint at what? And what are those stars doing there? These puzzles are supposed to stick to ordinary letters. You can't go adding stars to the alphabet (*and* allow them to stand for *different* digits—outrageous!) just because you can't find something between A and Z to suit your purpose. But Simon was too amused by the chase after numbers to worry about that. SIMON + NORTON = NO*HINT would be closer to the mark.

Simon can solve these puzzles with startling speed.

"Don't you have to think about it at all?"

"No."

"You just know it? You can see instantly that the letter N must be a 9?"

"Yes."

"It is somehow obvious to you at first glance that T will be an 8?"

"Yes."

"'Obvious' in a logical sense," I asked, "or 'obvious' in the sense of a sensation? Would you feel distressed as if you'd eaten a moldy tomato . . ."

"A raw one is disgusting enough."

". . . if someone suggested T might be 2, or 7?"

Simon looked perplexed. "This is a general phenomenon, isn't it? When one does something often enough one learns how to do it without conscious thought."

"But there's a difference. Do something quickly because practice has made calculation easy, or do something quickly because you don't need to calculate. The second is divination. That's genius."

"Huunnh," Simon grunted, and stabbed at his fish in distress.

* * *

I'd read once about a journalist who'd decided to learn the trick of calculating the day of the week, for any date.

"July 15, 1843?" you would ask.

He'd reply, "Saturday."

"December 30, 2076?"

"Wednesday."

"March 19, 12693 BC?"

The formula was easy. He memorized it in an evening. Daily practice made his speed increase. In a month or two he could answer any question in ten seconds. What took time was the arrival of fluency, the ability to think without thinking: to be able to do the calculation so readily that even to himself no calculation appeared to be involved. That took three years.

After that, for this journalist, October 12, 1646, was no longer a Friday because the numbers said so but because it could not conceivably be anything else. "Friday" was the only possible word that evoked the sensation of "October 12, 1646." It would be as foolish to mention "Tuesday" as it would be to put salt on your ice cream.

Is this what happened to Simon? Did his obsession with calculating two to the power of thirty when he was four years old rearrange the neurons in his brain (achieving in an afternoon what took the journalist three years—brains are so much more flexible at four years old) so that certain number problems became effortless after that—more a matter of sensation than calculation?

But this clumsy struggle to put my question into words had taken my focus off Simon. When I looked back at him now, he had put his plate to one side and was perched forward on his mattress, neck shrunk in, eyes staring straight ahead.

"Is that knocking?" he grunted. "Someone's at the door."

"Never mind that. We're homing in on the origin of genius. 'Sissors,' calculating all those twos and finding patterns, phrase-into-sum puzzles—as I see it, the essential elements are

a) playfulness, b) visual satisfaction, which would have made any attempt to force mathematical learning on you, like those ghastly aspirational parents, destructive by removing natural enjoyment, and c)—"

"The door!" Simon bounced up and down in frustration. The plate and cutlery clattered half a beat behind. "Someone *is* knocking!" he cried.

"—and c)" I pursued, "utter *self-centeredness*. Get the bloody door yourself."

16 Simon Cuttlefish

For neither at Milan, nor at Pavia, nor in Bologna, nor in France or Germany, have I ever found a man who could successfully controvert or dispute me within the last twenty-three years. Yet I do not vaunt my powers on this account, for I think that, had I been made of stone, the same things would have come to pass. It is the result of the lack of clear thinking on the part of those who would challenge me, and no more a special dispensation to my own nature or to my own distinction, than it can be counted glorious for the cuttlefish to eject the shadows of its inky humour about the dolphin and force it to flee; that is merely the result of being born a cuttlefish.

Girolamo Cardano, The Book of My Life *(1576)*

When Simon was three and a half, his mother arranged an IQ test.

In terms of IQ, "genius" is supposed to cover those with a score above 140, about a quarter of 1 percent of the population (one in every 400 people), although some researchers put the mark at 158, the top 0.13 percent, which is one in every 800. Taking the stricter figure, that means there are 80,000 geniuses in Britain. That alone should indicate what a pile of idiocy these tests are. According to these statistics every village the size of Six Mile Bottom has Shakespeare and Darwin slouching down the chippy for haddock and peas. IQ tests are good for predicting one thing only: success at school, and even in that they are not the best guide. The most successful measure of how a child will do in exams and future academic life is parental income. Most people who are successful and rich and happy and winning prizes for creativity and inventive brilliance don't have remarkable IQs.

But an IQ of 178 is eerie.

Such a person is not just very good; he is inconceivable. His aptitude is not beyond reach or comprehension; it is beyond description.

Simon's test lasted just over an hour.

At three and a half,d he could count from 0 to 100 in twos: ("Alex, a typo there! An extra 'd' on that last line.")

2, 4, 6, 8, 10, 12, 14, 16, 18, 20, 22, 24, 26, 28, 30, 32, 34, 36, 38, 40, 42, 44, 46, 48, 50, 52, 54 . . .

In fives:

5, 10, 15, 20, 25, 30, 35, 40, 45, 50, 55 . . .

And tens:

10, 20, 30, 40, 50, 60 . . .

And threes:

3, 6, 9, 12, 15, 18, 21, 24, 27, 30, 33, 36, 39, 42, 45, 48, 51, 54, 57, 60, 63, 66, 69, 72, 75, 78, 81, 84, 87, 90, 93, 96, 99.

He could read books, write out "write, right, wright" and explain the meaning of each. The woman conducting the test took him out for a walk as part of the warm-up routine, and when they came back he spelled out "fire extinguisher" (in cursive, because he was "tired" of print writing).

There was only one area in which he lost marks: "bead-threading," which involved stringing different-shaped objects onto a shoelace. She put square, circle, square, circle, then handed the string to Simon. "What's next?"

Blissfully, he chose triangle.

Gipsy Hill 3374

Dear Elaine,

I couldn't believe the Test score & phoned up my colleague to complain. Simon's Mental Age works out at 6 years which makes his I.Q. 178! I have never known such a result and have always been impressed with anything between 140 - 150.

I thought over our tea-time chat about Simon and I would like to emphasise the following points;-

1. To avoid mental strain & to achieve some sort of balance, could you encourage him to draw, paint, dance to music and play with one or two children? As a life-long Froebelian I cannot stress play strongly enough! — constructive play, make-believe, physical play with balls, ninepins, hoops, tops, skipping rope etc.

Letter from Simon's IQ examiner. Simon is arrowed in the photograph. On the other side of the letter, she reminds Helene (Simon's mother, whom she calls "Elaine") to keep the score to herself because it can "prove an unbearable strain even on the brightest child." The emphasis on playing is interesting. Simon has always treated mathematics as a form of play.

"Simon Norton? *The* Simon Norton? That's a blast from the past. Simon Norton—*here?*"

It's a good thing I did find out who was knocking at the end of the last chapter. It was new tenants for Simon's attic room. Shiny and Grubby, I named them. Standing in the living room (which, for reasons Simon cannot explain, is called the Ferret Room), they stared at me in a blur of brilliantine and bugs.

"I'd heard he'd gone mad," said Shiny.

"No," I corrected, "still quite sane."

"Crazy as a Torquay pavement. Memorizes bus timetables."

"Collects and reads them," I admitted. "Some of the information happens to stick in his brain, of course."

"Catastrophic mental collapse."

"Don't be ridiculous! He still goes to conferences and publishes papers."

"Of course, he'll never get it back now," said Shiny, with bright tragedy. "Mathematicians are over the hill by their mid-thirties."

"Except Fermat," pointed out Grubby. "He didn't start until he was forty."

"And Euler," agreed Shiny. "Going into his seventies."

"Like Littlewood."

"And Gauss."

"Archimedes."

"Cayley."

"Phew!" they exhaled. "But Simon *Norton. The* Simon Norton. He must be *at least* 100."

Shiny or Grubby? One of them had to get the attic room. These were the only two who'd showed up and not run away. Simon said it was my decision. I couldn't decide between them. They were both perched expectantly on the sofa, waiting for an answer.

Then I had an idea. I ran downstairs and brought Simon up from the Excavation.

"Set these two a math problem," I said. "They don't think you can hack it anymore. Give them a question to teach them a lesson."

Without hesitation, Simon snatched a piece of paper from the duffel dangling from his arm, scribbled quickly, tore the sheet in half and, grinning his grinniest, handed one portion to each of the two candidates.

It was a new phrase-to-sum puzzle, but this time it involved multiplication:

$$\frac{\text{SIMON} \times \text{P}}{\text{NORTON}}$$

Neither could do it—not on the spot.

But Shiny later phoned up to provide *two* solutions. Appalled, I gave the room to Grubby.

17

Symmetry is like a disease. Or, perhaps more accurately, it *is* a disease. At least in my case; I seem to have a bad case of it.

Joe Rosen, Symmetry Discovered: Concepts and Applications in Nature and Science *(1975)*

Mathematicians are subtle with language, but they have made a mess of the word "symmetry." What ordinary people mean by "symmetry" is kindergarten shapes (such as Square, Triangle, etc.) and snowflakes.

At the center of what mathematicians mean by the word "symmetry" is the word "always."

A square *always* looks the same when you rotate it through a quarter-turn, or half a turn, or three-quarters of a turn, or turn it backside up; or do any combination of two, five, ten, or 743 of these things. It is *symmetrical* with respect to these operations.

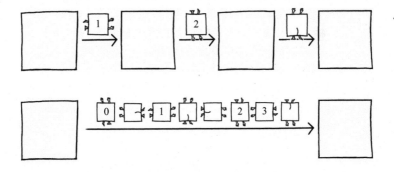

Triangle *always* looks the same when you manhandle it through a third of a turn, two-thirds of a turn, flip it bloomer-side and rotate it as many times as you like, or keep yourself churchlike and Sunday-mannered and pass it by altogether. It is *symmetrical* with respect to it all.

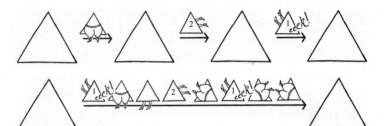

If you can phrase an idea—any idea, on any subject—to include the word "always," then you're within whistling distance of what mathematicians call "symmetry."

- Simon *always* fills his bath too full.
- I *always* get cross and kick shut the Excavation door when Simon cooks packet rice with Chinaman in it.
- E *always* equals mc^2.
- The Monster *always* looks the same (although no one yet knows what that look is like) when you perform on it one of the 808,017,424,794,512,875,886,459,904,961, 710,757,005,754,368,000,000,000 symmetry operations in its Group Table.
- When drunk, Jim, our next-door neighbor, *always* leaps on his motorbike at 2 a.m. and starts revving the engine.

Each of these cases makes an appeal to symmetry. Symmetry is the invariability of some object, circumstance or relationship to a specified set of changes. Put Simon in the bath in our house or on a planet in Alpha Centauri; in the middle of a nuclear war; when he was five, seventy-five or upside down—it makes no difference. As far as the water in the bath is concerned, it'll *always* dribble over the edge, seep under the skirting, ooze along the basement wall and cause toilets and stair treads to collapse. It's a knack Simon has.

In the starched language of squares and triangles, "rotate" Simon-in-a-bath among any of these different situations, and the result always looks the same:

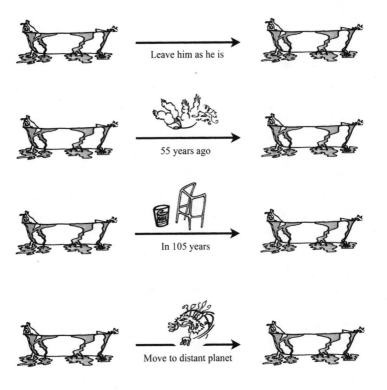

Leave him as he is

55 years ago

In 105 years

Move to distant planet

However many times, in whatever order, we perform these operations, it will still be splish-splash Norton in the bath:

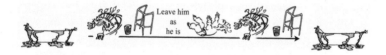

The anthropologist Claude Lévi-Strauss noted that certain types of marriage among Australian tribes are *always* considered incestuous and therefore forbidden. But he couldn't figure out the underlying structure behind these taboos, so he went around New York (where he was living at the time) banging on the doors of mathematicians. The first was dismissive: "Mathematics has four operations, and marriage is not one of them." But the second was the young and brilliant André Weil, brother of the philosopher Simone. "When in doubt," cried Mr. Weil, "look for the group!" and he bustled Lévi-Strauss off the street into his study. Within a few days, Weil had solved the problem. Not only had he used the theory of symmetry to explain why certain marriages were taboo among Australian Aboriginals, but he'd also discovered that, by mathematics alone, it was possible to investigate a tribe today and determine if at some time in the past it had been in fact two different tribes, which had since met and settled together. There would be an imperfection in the "symmetry" of the incest taboos—the "alwaysness" of their rules—that would give the game away. You didn't need to know the history of the tribe, or have any further anthropological knowledge than an understanding of the current restrictions on sex: all you needed was to pick up a glint of asymmetry, the murmur of a "sometimes." With mathematics, you could see 1,000 years into the past.

I've charged about my study for hours trying to understand how going to bed with your sister is a mathematical statement, and failed, but it is.

("I don't understand it either," grumbles Simon. "In fact, I'm not sure I believe in it.")

The word "taboo" suggests *always* forbidden, so symmetry must be involved somehow. "Rotate" any one of the banned couplings in front of Australian tribal elders and there will *always* be uproar.

This is why symmetry is important: it's *everywhere.* Symmetry does not mean just preschool cut-out shapes like triangles and squares, it speaks for anything—from quantum fluctuations inside your eyelash, to the songs of the Beluga whale, to the despotic delusions of madmen in North Korea— that involves, in some respect, however far-fetched, "immunity from change." It is the study of imperturbability. If, under certain conditions, a situation *always* appears the same despite the fact that things are actually being done to it, then that's symmetry.

This tells us something about Simon's Monster. It is not a raging thing. Its power lies in its unchangingness. It is a monster of calm.

And here's a fifteenth way to say it: a symmetry operation is an act that leaves a thing "unfazed." Mess and boring bus journeys, for example, leave Simon unfazed. Mess and buses are a part of Simon's symmetry.

"I don't think that's an analogy at all," interrupts Simon, looking up, tongue clamped between teeth.

"*Loosely,* 'being unfazed' is an analogy to being symmetrical," I pursue.

"No, *not* loosely. The two ideas have nothing to do with each other. A symmetry operation in mathematics is something that *operates* on something else. It needs to *do* something to be interesting, as in a rotation *operating* on a square. If that rotation makes the square appear different than it did before, it is not a symmetry operation; if the same, it is."

"Exactly. Operate on you with mess, and your character and mood don't change."

"What do you mean, 'operate' with mess? Operate how? I don't understand that at all. The character of a human has nothing to do with operations. It's just to do with what a human is. It's nonsense to say they're analogous. I don't even know what that would mean, to talk about the mathematical character of humans. I really don't want to talk about this anymore."

Mess, buses and *nitpicking* are part of Simon's symmetry.

18

There are types of authors who are not O.K., names it is O.K.
to pitch into. It is alright to pitch into:
Any author who has written a book about dogs.
Any author who has written a book on natural history,
illustrated with woodcuts.
Any author who has written a life of Napoleon, Byron, or Dr.
Johnson, without footnotes or bibliography.
Any author of a life of anybody not yet dead.
Any author of a book on Sussex.

Stephen Potter, Some Notes on Lifemanship *(1953)*

"Did I tell you," declaimed Simon in a buffalo voice, "when I was a boy I could pee for longer than anyone else?"

Our train was heading through the chalklands of Sussex toward Ashdown House Junior School. It was one of those over-eager electric services that shouts at you to stop smoking and refuses to let you be in charge of the lavatory door.

Me? I was over-eager too. I'd had a biographical smash hit! The third person sitting with us, a tall, thick-set man with fingers chunky as carrots, was Malcolm Russell (not his real name—he wants to keep that secret), one of Simon's old school bullies. He'd Googled Simon, then emailed him. Tormented by what he'd done half a century ago, he wanted to . . . to what? He wasn't quite sure. To apologize? To make amends? To understand? Not knowing which, he'd proposed this trip back to the school instead.

Already, his influence had transformed Simon. Ever since East Croydon station, Grunter had become Babbler. It was astounding. It was miraculous.

"I could stand further back from the urinal too," Simon hollered again, filling the carriage with triumph.

It was embarrassing.

I turned to look out the window in shame. Purple and white crocuses clustered along the banks, and pastures were puddled with water. Oak trees had begun to bud in the cold, sun-startled air; rabbits, trembling with winter, crouched among daffodils. Everything was precarious and young, and two months early, chancing on there being no frost in April.

Malcolm runs a fertilizer store in Shropshire. After Ashdown, he'd gone to Radley, a public school next to Oxford. Memory has scraped that hellhole deep into his conscience. "Other boys had mental breakdowns while I was there. One committed suicide."

Malcolm is constantly searching his pockets for not-too-discomforting recollections of childhood.

"The urinals!" he roared now with delight, sinking his hand into a packet of salt-and-vinegar crisps, exploding the bag. "Yes, I remember the school urinals! I can see the loo. I remember it exactly. I really hope they haven't pulled them down. We'll *have* to go there. We'll have to see *every*thing! And the Spanish teacher, Mr. Juh—, Juh—"

"Jabuzi!" cried Simon.*

"Yes! Did he put his hand up your shorts too?"

The school announces it is "set in rolling hills," but it isn't. The Downs roll. Devonshire rolls. North Sussex wobbles. William Cobbett called the glorious Sussex countryside around Ashdown "the most villainously ugly spot I ever saw in England."

---

*"I don't remember there being a Spanish teacher at Ashdown," protests Simon. "Of course not, Simon. I can't say the right name and what he really taught, or it would be libel."

Seed-speckled haze; woodpeckers scooping between the spruce trees; cuckoo cadence lilting in the breeze: Ashdown House is set at the end of a half-mile drive in forty acres of rhododendrons. After the wood (which starts to the right of the school and is all that remains of the ancient Ashdown Forest) comes the high, scratchy heath, then a stretch of gentleness much more to Cobbett's farmhouse taste, followed at last by the escarpment to the South Downs.

The school charges £5,400 a term for non-boarders, and unmentionable sums for boarders. It has alumni who are mayors of London (Boris Johnson), have been sent to prison for shooting poachers on their estates in Kenya (The Honorable Thomas Patrick Gilbert Cholmondeley of Delamere) and write novels about Notting Hill Yummy Mummies (Boris's sister Rachel). As a schoolgirl, Rachel found a "live, plump maggot wriggling" in her shepherd's pie. She bore the plate "with its live cargo" across the dining hall and pointed out the ghastly object to Clive Williams, the headmaster.

"Clive peered at it expressionlessly. 'Yes, very nutritious, I expect,'" he said, and dismissed her.*

In the cattle-farming language of the wealthy, Ashdown is known as a "feeder" prep school for Eton. That's why Simon's father (or was it his mother?) wanted him to go there. It's the best school for getting boys into Eton. Eton is the best school for getting boys into Trinity College, Cambridge. Trinity is the best college for providing genius mathematicians with a pleasant life.

The main school building was designed by the architect Benjamin Latrobe. After completing it he migrated to Virginia,

*Rachel Johnson quoted in *The Hill,* Issue 287, September 2008, p. 66.

where he designed the State Penitentiary, taking a particular interest in solitary confinement.

"To be honest," admitted Malcolm quietly as we walked up the drive among the cuckoos and seeds, Simon yards in front of us, "when I first emailed Simon last month I half thought he'd be banging his head in an institution or have died, though not suicide—car accident, I expected.

"There was a kind of sweetness to him when you sat down and paid attention to him. There was a sense of him being defenseless in the world. You really felt him soften, he liked friendliness so much. If you asked him who the Prime Minister was, he might say someone from the First World War: 'Asquith.'

"He certainly didn't make you envious of genius," he added. "To be like Simon seemed a high price to pay."

A cluster of pigtailed girls poking out their tongues and waving hockey sticks rushed from the school gates ahead, pushed us off the verge of the drive and, hopping over a hump in the road, surged toward Simon.

"Once—do you remember this, Simon?" Malcolm called to Simon, who was making miserable efforts to regain his balance following the pigtail attack. "—There was a school quiz and you were asked to spell 'bikini.'"

Simon chuckled contentedly.

"Who else in the world but Simon would you know who spelled it 'Becquigny'? The only thing he could think of that sounded like 'bikini' was the Treaty of Becquigny between England and France in 1340."

"Pecquigny," interjected Simon, shocked by the scent of a false fact.

"Becquigny," repeated Malcolm.

"Huunh! No! *Pecquigny*. August 29th, 1475. Louis XI paid Edward IV 75,000 crowns and had to ransom his wife, Margaret of Anjou."

Malcolm shrugged his shoulders, stepped back onto the drive and sighed happily. "I've talked about him a lot over the years," he confessed.

A car zipped past us and coughed to a stop in the car park.

"Excuse me, can I help?" exclaimed the driver, bounding out onto the gravel and striding over. He gave a vicarish peer.

"Well! If it isn't Simon Norton! Ashdown's greatest scholar!"

Clive Williams, ex-headmaster, had come back for the day. A former Ashdown boy, he'd spent five decades at the school and, when he'd got wind that Simon was visiting, had raced back to it from his house in Lewes to say hello. Even after forty-two years, he recognized in a second his old fellow pupil.

I'd recently interviewed Clive about Simon's time at Ashdown: four hours sitting in Clive's cozy kitchen, going through the school bulletin, him gazing at me and the middle distance while he remembered the golden schooldays of the 1950s. But I am not an Ashdown boy. I am, in the kindest sense, dust to a man like Clive. After greeting Malcolm with more cries of happiness, he now turned to me and held out his hand:

"And who," he said pleasantly, "are you?"

It's the subjects that Simon couldn't do that are interesting. In mathematics, of course, he understood everything he was taught, and divined the rest.

In Greek and Latin (as so often with mathematicians) he was a vacuum cleaner.

To a certain extent, this is down to memory. A mathematician who can't remember obscure facts is, like a detective who can't remember clues, impossible. Mathematical ability depends on being able to make links between remote ideas, and people such as Simon never do this by the Inspector Plod method of dully applying and reapplying the rules of school algebra. They sense the correct approach, sniff it out like a dog. This doesn't mean they've just got more sprightly than normal nerve endings in the mathematical parts of the brain; it means they have a superb memory for certain types of details, can use it to draw comparisons with other mathematical discoveries they've made or read about elsewhere, and therefore can exploit tiny and hidden analogies of argument.

If Simon had not become a mathematician, he would have made a very good scholar of dead grammars, publishing compendious books that were instantly outdated.

Yet in history, a subject you'd think would appeal to his excellent memory and obsession with fact gathering and numbers, Simon was at sea. "Sublimely indifferent," according to his school reports, his essay on Warwick "extremely feeble":

"I could never understand what history was about," says Simon. "Why were they always fighting over a field?"

"But you remember the Treaty of Pecquigny."

Religious education was the same, despite the fact that Simon's father was president of the Jewish Reform Synagogue, and the Ashdown headmaster—a brilliant and charismatic scholar called Billy Williamson—used to pace through the classrooms each morning, Bible in hand, reading passages of gore and genealogy in a booming voice that carried across the entire school.

Warwick

Richard Neville, Earl of Warwick was a supporter of Richard of York. He was known as the King-maker because he crowned two kings, Edward IV in 1461, and Henry VI, who had been king from 1422 to 1461, in 1470. When Henry VI died in 1471, Edward IV again became king, but this was nothing to do with Warwick, who had been killed at the ~~Battle of Barnet~~.

In 1455 Warwick was ~~the~~ ~~...~~

Written after his pen had fallen nib-first on the floor.

Billy Williamson operated a five-year plan. By the end of each five-year stretch—the length of time a boy was at the school—he had declaimed the whole of both Testaments. If you arrived for your first year when he was starting Genesis, you left for Eton or Harrow at the end of Revelation. If you appeared at the Second Book of Kings, with the little boys who persecuted Elisha being eaten by bears, you left, sixty months later, as the headmaster read from the First Book of Kings about randy King Solomon and his 700 wives and 300 concubines.

Williamson did his best to make the subject engaging to odd pupils such as Simon who didn't love the usual bloodshed and despicable behavior of our God in the Old Testament. In an attempt to nudge Simon at least a step or two into an interest in the history of his own people, the headmaster included this question in an end-of-term test paper:

If you complete the following sum correctly you will arrive at a critical date in Jewish History. What was it? Multiply the number of Jacob's children by the number of Zeruiah's sons and then by the number of brothers left by Dives. Then subtract the date of the destruction of the Samaritan Temple on Mt. Gerizim by John Hyrcanus, plus the number of tribes of Judah multiplied by the number of tribes of Israel after Rehoboam's reign.*

Simon was no better at this than he was at the Earl of Warwick and his wretched king making. Yet what Billy Williamson may have done successfully was to encourage Simon's mental playfulness. He used to let him lie about on the floor of his study for hours, doing sums and laughing and enjoying being teased and playing word games.

"Simon," wrote Williamson in one of his delightful end-of-year reports, "goes on his way rejoicing, and I rejoice with him."

It barely needs saying: sport was Simon's worst subject. This will be a short section.

"During a game of cricket he spends his time counting blades of grass or calculating angles. He takes about as much interest in the proceedings as Archimedes did during the siege of Syracuse." In football "he flutters about like a butterfly." During swimming, he sank; for rugby, he didn't move.

"I used to stand on the pitch," admits Simon, "and eat the sorrel."

* * *

---

*I can't make this come out, although I've worked at it all afternoon. Jacob had thirteen children; Zeruiah, the sister of King David, had three sons. Dives was the rich man who refused to give Lazarus crumbs from his table: he went to hell and left behind five brothers. The destruction of the temple occurred in 128 BC. The tribes of Judah were one. It *is* a tribe. It split from the rest of Israel after Rehoboam's reign, and set up camp in the south with the tribe of Benjamin, leaving behind the remaining ten tribes.

$$13 \times 3 \times 5 - (-128) + (1 \times 10) = 333$$

333? What's critical about 333 AD in Jewish history?

The question, agrees Williamson, is "beyond the capabilities of most, I really don't know why" (*Ashdown House Bulletin,* December 1964, p. 29).

In other subjects, Simon's failure required more effort. It takes determination for a boy to be at the bottom of the class in geography. Simon sank there like sedimentary graphite. Incapable of disguising boredom, he'd rather endure hours of disapproval than five minutes of homework about why rivers wiggle or the post-war output of a Dagenham car factory. This is a limitation that would crop up later, in his work as a mathematician: if he doesn't take to a subject and feel excited about it and want to dominate it immediately, he can't be bothered with the thing. Intellectually, he is extremely lazy.

I doubt whether I tried as hard as I could

Simon acknowledged in a report he was asked to write about himself one term, when the geography teacher was off sick.

my map-drawing is appalling

"In some ways he is extremely competent," conceded the English teacher: "grammar, syntax, spelling & now, even writing. Yet there are definite deficiencies. He has very little general information at his disposal; he neither reads aloud nor recites well; yet he can remember any poem he learns with the greatest ease . . . He still stoutly refuses to admit that he has enjoyed a book." His written work was "obscure to everyone but himself."

E G I E may

Is learning dangerous?

I am afraid that I find this totally incomprehensible

Is learning dangerous? This is one of the debatable questions where the answer is not necessarily yes or no, but sometimes, often, usually, or some other such adverb of frequency. All rules have exceptions (except, of course, this one!) because if only a little is learnt, the rules in this case, the exceptions are not recognized, but, of course, a little learning is better than no learning, because an exception to a rule is only one case, and if there are too many what was a rule cannot be a rule after the addition of more. The borderline between rule and no rule is hard to establish.

I am inclined to agree! →

This is digressing and, to return to the point, a little learning is better than none, to a greater extent than that to which much learning is better than a little, for otherwise the rule, which makes the learning of only the rule dangerous for the case of exceptions would not be a rule, and as this may well be the critical question, the answer to this question is in favour of the negative.

On the other hand, there is another way of looking at the question and its supporters would answer in favour of the positive. A little learning must be, to some extent, dangerous, for anything in which even the tiniest bit is missing is to some extent dangerous, and someone might equally say that living in a house is dangerous, as there might be an earthquake. Therefore, such a person might say, it is better to live in primitive tubs.

I feel that, somewhere in all this, there must be some sense, but I am afraid that I have not been able to unearth it.

Transcription from the handwritten original. Simon was ten. Discovered under the bedside table in the Excavation, forty-seven years later.

As the shock of Simon's precocity wore off, a note of exasperation crept into his teachers' comments. "He has literally danced his way through, much to the chagrin of his form-mates," wrote the French master at first, when Simon was still able to get by on just his memory, but he adjusted his views soon after. "He has the most annoying/amazing habit of getting difficult work correct while he spoils it all by making several elementary blunders." "He knows his grammar thoroughly [but] he is so lacking in imagination that his essays & free compositions are often meaningless."

In Greek there was a suspicion by the teacher that in this subject Simon might be extra-terrestrial, but "he makes just the right number of mistakes in his work to show that he is human."

If the subject could be reduced to rules, Simon invaded. If it couldn't, as in geography or history or scripture, he doodled in his notebook and started on impossible sums:

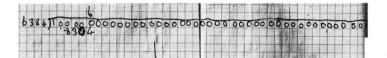

Aged ten, he triumphed also at music, and wrote a sonata, 100 bars long and pronounced, forty-five years later, by internationally renowned harpsichordist Giulia Nuti at the world premiere in her front room in Florence, "a very peculiar piece of music indeed."

"Do you remember Collins?" Malcolm whispered to Simon, as Clive took us upstairs to see the school dormitories. "Did he also put his hand up your shorts and pinch you?"

"No," retorted Simon in a loud voice. "He didn't do it to me. Maybe it was because I was good at playing the piano."

Simon's only continuous error while at Ashdown House was to keep dropping his pen on the floor. But that didn't stop him if it happened during a subject with rules. *Write, write, write* . . . he raced, unstoppable, ink splashing . . . *write, write, write* . . . wrenched, twisted half-nib digging through the pages of his exercise book . . . *write, write, write* . . .

"The results are fantastic," cried one teacher in despair, angling and twisting these deranged, splattered pieces of classwork in the light.

"We have failed—partially failed, anyway—to teach him to live as a component part of a community," sighed the headmaster, then added: "I wish he would dress, feed and sneeze more tidily."

Simon was put in a cage.

The headmaster ordered it.

"A metal cage?" I asked excitedly. "Like a canary? What a fabulous idea!"

It was not a metal cage. It was a cage of books and cushions built up around Simon during his Greek lessons and called, for reasons no one appears to remember, his "Bruton Cage."

Billy Williamson had noticed how much Simon enjoyed isolation and compact spaces, so he reasoned that if the peculiar boy was encased entirely in a grotesque mountain of paper and soft furnishings he would feel blissfully alone.

The smallest things bothered Simon. At that age (seven to twelve) he lived in a billow of superstitions. In the dining room, he would throw a fit if he spotted a knife pointing in his direction (so, naturally, the boys arranged theirs accordingly). He worshipped the number seven.

"One day," recollected Clive as, with a church-bell swing of his hand, he scooped us through the front door of the school into the lobby, "Simon came down to breakfast frightfully excited and said he'd just invented a new system of multiplication tables based on seven instead of ten. Nobody knew what he was talking about."

Simon doesn't either. "Hnnn, aaah, no. I don't remember that at all."

"Simon," I exclaimed, stunned, "are you saying you re-rediscovered modular arithmetic when you were at prep school?"

"That's exactly what I'm *not* saying. I'm saying I don't remember."

Revealed by an unknown genius in the East, in the West "modular" or "clock" arithmetic was rediscovered for the first time by the teenage Carl Friedrich Gauss, the "Prince of Mathematicians." This was regarded as one of the great achievements of the eighteenth century. "That makes you equivalent to Gauss," I squeaked, in awed falsetto.

The mention of the holy name, Gauss, brought Simon back to the land of good temper. "As I say, I was not a Gauss. I was certainly never a Gauss. No, no, heh, heh, huunh. Not a Gauss. I think I'd like to use the toilet now. I'm full of liquid."

"He was like a toy," said Malcolm, relaxing a little with these memories of Simon's boyhood peculiarity. "It wasn't ordinary bullying. It was almost communication. If you were standing next to him in the urinals and made a fart noise with your mouth, he'd bounce up and down, all the way down the wall, spraying urine everywhere. It was rather fun."

"But when he was in the cage, he was safe?"

"For a few minutes, yes."

"And then?"

"Isn't it obvious? Somebody made a fart noise with their mouth."

All the same, none of this persecution seems very serious—certainly not serious enough to justify Simon saying that the boys here ruined his self-confidence for life. When the Volga Bulgars see a prodigy, they say, "It is fitting for this man that he should serve our Lord," then seize him, fling a rope around his neck and hang him from a tree until he disintegrates. That's proper bullying.

In his forties, Malcolm had attended a Boarding School Survivors workshop to try to come to terms with the bullying he'd endured at Radley. The *Wall Street Journal* ran a front-page article about him. He appeared on an Esther Rantzen program about public schools. The youngest boy at the workshop had been nineteen; at Eton, he'd been dumped in a bath full of shit and piss. *That* is *real* bullying.

And to some extent Simon deserved it. Doesn't he himself admit that one of the things he did in his spare time was to "reorganize" the school timetable so that "it would be more efficient and not include so much unnecessary time between lessons"?

"There were other bullying things too," acknowledged Malcolm, with a guilty groan, as Simon returned from the toilet ("There, now I can start filling up again") and pattered ahead, up the staircase, toward the dormitories and classrooms.

"What things?"

Malcolm shook his head in dismay.

"Malcolm! *What* things?"

We arrived at the attic room where Greek was taught. Simon hurried across to feast on a map, grabbing his arm behind his back in delighted concentration.

"No. It's better to let sleeping dogs lie."

"*Malcolm, spit it out!*"

Malcolm peered after Simon, looked at me; peered back at Simon; sucked his lips hesitantly, then whispered quickly:

"We called him . . ."

19

"Cab—AB—AB—ABB—Baabbb—baaaaaa—gggge!"

"Don't." Ten-year-old Simon: face blank, eyes scuttled.

"Cabbuuu-uuu-UUU-AAAAAAAA-ge," repeated boys at Ashdown.

"Don't," replied Simon in an unapproachable drone,

> smile stapled to cheeks.

"Oi, *you*, Cabbage!"
"Don't."

"Cabbage, Cabbage, Cabbage!"

"Don't, don't, don't," repeated Schoolboy Simon.

"It was like ping-pong," says Malcolm. "You'd say 'Cabbage'—"

"Don't."

"—fifty times, and he'd say 'Don't' fifty times, always in this same affectless voice."

"C A B B A G E !" bullies
yelled across the fields
during rugby matches,
when Simon stopped by
"Don't," said Simon   to eat sprigs of sorrel out
not raising his voice,   of the grass.
between chews.

"C A B B A G E !" they shouted as he
wandered off to sit in a rhododendron bush, with his chess set
and university textbook on complex geometry.

"Don't," said Simon.

"Cabbage! Cabbage! Cabbage!" clanked the bell in the chapel tower.

"cabbagecabbagecabbage" pattered schoolboy feet rushing to school prayers.

"Cabbage?" called the dinner ladies. "Cabbage!" roared the school.

"cabbash, cabbash, cabbash" squished their teeth.

"Don't," said Simon.

"Cabbage, cubbige, cuhbage, cabbage, caba-caba-cabacabbage," they

whispered into his ear when they thought he was

ASLEEP

"Don't," said Simon, opening his eyes.

"Carrot."

"Phew. Thank you."

"CABBAGE!"

"Don't."

"Sauerkraut."

"Not in any language."

"Cabbage, Sauerkraut, Cabbage, Sauerkraut, *Chou* . . ."

"Don't," said Simon.

"*Chou*. Shoe, ," gesticulated the bullies, grabbing up the pictured object.

"Don't."

"Cabbage, Sauerkraut, *Chou*, , . . ."

No one understood why "cabbage" . . .

"Don't."

. . . upset him so much, in any one of five languages: French, German, English, Latin and Homeric Greek.

"Do you know what I think?" I said as we left the classroom and I brought the subject up with Simon.

("Don't!" said Malcolm. "I really *don't* think you should.")

"I think it's nothing to do with cabbages. It's just that the word 'cabbage' somehow, in your mind, came to stand for bullying. When somebody said 'cabbage' to you, you knew what he was actually saying was, 'I'm bullying you.'"

It's like x in an algebraic equation. It was shorthand for "I'm too bored to do actual physical bullying in this circumstance: let 'cabbage' stand in for whatever that bullying would have been, were I to have done it." The language didn't matter, because bullying's bullying wherever you go.

I was pleased with this notion of algebraic bullying, and as the four of us creaked along the shadowed floorboards to the girls' dormitories (named "Boudica," "Pankhurst" and "Margaret Thatcher") then back out of the building and into the limpid light of the playing fields, I rephrased it to Simon several times.

"It's physical algebra . . . Like money too, really, when you think about it . . . £20 stands for any purchase, be it three bottles of wine, a pair of shoes . . . the currency of bullying . . . the currency of algebra . . ."

But Simon and I never seem to agree on the depth of my insights, on any subject. As we stepped onto the rugby pitch—goalposts padded to reduce collarbone breakage, grass sprinkled with schoolboy leg flesh—his eyes glazed over and he heaved a sigh, which wasn't bored or upset, but more like the noise a train emits after a long rush between stations: a result of too much air inside the brakes. "Oh dear!"

Then he perked up. "Last week I went to Spinach, Shropshire. Would you like to see the ticket?"

"I don't think 'brilliant' is the right word," whispered Clive. We were now ambling past the cricket green, each of us with our hands behind our backs in a contemplative and umpireish manner, although Simon had somehow strayed off and got himself entangled in a bit of rhododendron bush. "Not for the non-mathematical subjects. It was a filing system he had in his mind. He was told something, registered it, reproduced it. But he *was* odd. He was just so out of kilter with anything we'd ever come across. You just couldn't label him, other than by using the word 'genius,' which, of course, a good headmaster very, very rarely used. To go around boasting that you've got a genius in the school is not good for anybody."

I remembered the French teacher's comment that Simon lacked "imagination," and I thought again that it wasn't quite fair. How could a boy who would grow up to stun the mathematical world with his capacity to explore the fundamental mysteries of the universe lack "imagination"? What was it, if not "imagination," that enabled a ten-year-old to see instantly (like a painter effortlessly spotting the essential lines of a composition, or a poet picking out resonant metaphors and instinctively disregarding tired and muddied alternatives) five unprecedented ways to arrive at a solution to an abstract problem that the math teacher, with all his step-by-step traditional "workings," couldn't clunk through in a year? Who but someone with the "imagination" of a da Vinci would one day understand and describe objects that existed only in 196,883 dimensions?

It wasn't "imagination" Simon lacked. It was French that lacked standards and constancy. When Simon began at Ashdown, the teacher had suggested that the essence of communication in French consisted in learning things such as irregular verbs, past and future tenses, the declension of pronouns; and that by the calm and regulated application of these one eventually arrived at fluency. These components sparkled with metallic certitude. But it turned out not to be enough to build up language from such reliable things. The French teacher then changed tack and demanded some of that "imagination" thrown in too—stuff that took an accurate and succinct sentence, correct in all its grammatical particulars, as per the appendices at the back of the book, and turned it into a load of jellyish nonsense that was even more lame than the math teacher's "workings."

Simon's trouble was that he had a tight aesthetic. He wanted to be right. He found *correctness* beautiful. Certitude and universality, not floppy talk about sunsets on the Côte d'Azur, were what provoked his "imagination."

"I developed the theory," barked Simon, suddenly behind us, cutting off the sunlight, pieces of rhododendron poking out of his hair, "that when someone punched me, if I couldn't hit the person back, I could do equally well by hitting someone else. Then they could settle it amongst themselves."

"But this is perfect! We've had the algebraic view of bullying, now, this is the arithmetical type. Your view was that it didn't matter which bully got thwacked, just so long as the *overall* effect was to put the punch back in the bullies' camp."

"I suppose so," mumbled Simon, suspicious that I seemed to be making sense so far.

"Which is the same situation as you get in arithmetic sums." Take, for example, the sum:

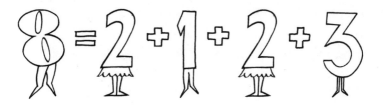

Now add 1 to each side:

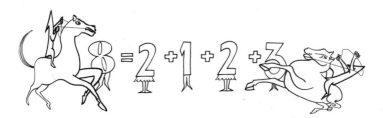

The left-hand side is obvious. It goes up from eight to nine. But which of the digits on the right-hand side gets the one added to it? The three? One of the twos? The one?

Oh dear, oh dear.

The correct answer is, it doesn't make any difference. All that matters is the *overall* effect, not this silly fussing.

"It's the same with you and your punches. There's you on one side, your bullies on the other:

"You get a punch:

"Which means they need a thwack back. But they're all bullies, so it doesn't matter which boy you punch back, whether he's the fat patsy or the knife-throwing son of a Mafia don. Just as with sums, it's the *overall* effect, not silly fuss about details that you're after:

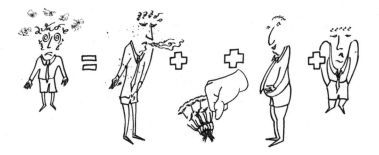

"You know what your trouble was, Simon?" I said, as we stepped back into the building out of the sun. "You got humans and numbers confused."

Despite Simon's lack of interest in this new idea, I think there's a profound point in it. Simon at Ashdown had an unbalanced mind. His mathematical genius was impossible for ordinary brains to comprehend, even the brains of adult mathematics teachers. He had raised mathematics out of logic

163

into artistry. His instant eye for the balance and elegance of a solution—I don't know how better to define it, never having got close to experiencing this myself—was a more effective judge of correctness than the inviolable classroom "workings." In short, for Schoolboy Simon, mathematics was not a separate subject—it was an aesthetic and a yardstick. You did not "do" mathematics between 11:00 and 12:30 on a Monday and a Thursday, any more than you did "admiration" or "fun." Mathematics was simply there: the setting for existence; the touchstone for all activity. Mathematics was to Simon what green fields and dark woods were to other schoolboys: the enjoyable places that you rushed to, whooping, as soon as the afternoon bell rang, to have conker fights and muddy your knees. Simon's little joke about hitting bullies was connected to arithmetic not because Simon was good at a distinct subject called "mathematics" but because mathematics was everywhere, and everywhere was mathematics, and for a few moments, before Simon found himself being repeatedly stabbed in the flowerbeds by the Mafia don's son, it had been good enough to let him share in its universality.

("I think we could insert a campaign reference, here, Alex. You could have me say: 'Come to think of it, arithmetic bullying is what the government is doing with public services. The bankers rifle our savings, the government has to return the money, and Cameron insists on getting it by axing key public services, including buses. I hope I didn't give the government the idea unwittingly because Boris picked up a bit of Ashdown folklore.'"

"You mean Boris Johnson, Mayor of London, friend of David Cameron and an ex-Ashdown pupil, might have passed on the school folklore about your method of dealing with bullies to the Prime Minister, and that would have given him the idea of punishing buses for the bankers' greed and mistakes?"

"I mean it as a joke, of course. Perhaps you should include: 'Simon added jocularly.'")

Simon had a second tactic *re* punches. It was cleverer. He showed love.

Instead of punching his persecutor back, Simon would open his arms and attempt to embrace the boy.

He knew exactly what the effect was. "I started as soon as I realized that hugging had a sexual connotation."

It scared the daylights out of everybody.

"On February 4th the Queen Mother had a successful appendix operation and Miss Beaumont in sympathy retired to bed." "On February 7th the first yellow crocuses appeared." "On February 10th Simon Crane produced mumps." "Just this side of Honiton a woman, in a small car, pounced out on me from a filling station and contrived to ram the rear offside mudguard of my car . . ."

At the end of each term, Billy Williamson, the headmaster, typed up the school *Bulletin*.

"On May 17th came the sad news of the death of Lord Brabazon of Tara, one of the great pioneers of motoring, aviation and tobogganing." The "lovely weather at Whitsun" was "sadly marred" by the "necessary arrest of 76 Mods and Rockers at Brighton." Brighton and Lord Brabazon had nothing to do with the school: the headmaster just liked cars and disliked scooters. "Next term Norton will try for Eton."

As a practice, Simon did the Charterhouse School math paper, for which two hours were given, did *all* the sums in twenty-five minutes and gained 100 percent. He tried the Radley School scholarship exams, did *all* questions on all three papers, well within time, and again scored 100 percent for each. While he was in bed with chickenpox he did the three papers set for the Assistant Masters' math prize at Eton, "got everything right and was easily top."

For the Eton scholarship exam, Billy Williamson wrote ahead to ask the setters to give Simon a really beastly question, one that he "could get his teeth into"—something from the Cambridge University entrance paper, for example. "Oh, don't worry," wrote back the Eton people (who had heard all the rumors about the young genius), "we certainly will."

"For logical thought, analysis and visual memory," wrote Williamson in the final issue of the *Bulletin* dealing with Simon's days at the school, "I don't suppose there will ever be anyone quite like him here again . . . He is unique. I am told that at the age of three he had an IQ of 85."

Williamson had a Jilly Cooperish attitude to any test not involving biblical history or Pindaric odes. He meant, of course, 185.

Two days before Simon was due to sit the Eton paper, disaster: gastric flu struck the school. "Twenty-nine boys, three masters and one under-matron in bed," Williamson informed the *Bulletin*.

Simon was dispatched to his grandmother in Woking for safety, and Billy Williamson followed shortly after, inching down the flooded roads through a storm to meet Simon and the other candidates at the Eton examination hall. "Lurid flashes of lightning were followed at ominously short intervals by loud crashes of thunder." For the next four days it rained continuously, with "only one small mackintosh and one umbrella between us." But through the downpour, out of the examination room, filling the spring air, they heard a boy's voice . . .

"Sir, what's that?" asked one of Williamson's companions.

"What?"

"That strange yodeling noise, sir."

"That," said Williamson, putting his ear to the wind to pick out the notes, "is Norton, in his mathematics exam, singing for joy."

Thank God for Porter!

Porter was another Ashdown boy, sitting the same scholarship entrance paper. At his preliminary interview, the Eton beaks had demanded:

"Why should *you* be allowed to come to the school?"

"Why, sir," said Porter, "to look after Norton."

Porter knew exactly how to shut Simon up in the exam and get him back to his desk:

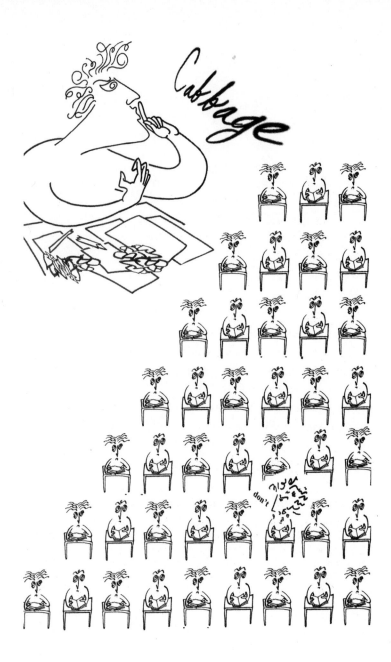

20 Eton

These have been the three happiest days of my life: an infinite variety of food, a minimum of exercise and a maximum of fascinating problems.

Simon, on rushing out of the Eton exam hall

From the age of twelve, Simon starts to become incomprehensible to me.

At Ashdown House Junior School, Simon worked with infinite series, negative numbers, modular arithmetic and "imaginary" numbers. I grappled with those things at university: they are within grasp.

Infinite series are sequences of numbers or letters that go on forever, according to a pattern that is always a little bit different from the one you suggest.

Negative numbers are understood by anyone outside of Manchester.*

---

*From the *Manchester Evening News*, November 3, 2007:

A LOTTERY scratchcard has been withdrawn from sale by Camelot—because players couldn't understand it. The Cool Cash game—launched on Monday—was taken out of shops yesterday after some players failed to grasp whether or not they had won. To qualify for a prize, users had to scratch away a window to reveal a temperature lower than the figure displayed on each card. As the game had a winter theme, the temperature was usually below freezing.

But the concept of comparing negative numbers proved too difficult for some . . . Tina Farrell, from Levenshulme, called Camelot after failing to win with several cards. The 23-year-old, who said she had left school without a maths GCSE, said: "On one of my cards it said I had to find temperatures lower than –8. The numbers I uncovered were –6 and –7 so I thought I had

Modular arithmetic (also called "clock arithmetic") is what we use every day to tell the time with a twelve-hour clock.

Imaginary numbers have nothing—at least at first—to do with the imagination. Ah, shame. They are stolid, practical little beasts, good for describing circular motion. They appear repeatedly in turgid university lectures on electricity and magnetism, because both of these forces depend on electrons, and electrons fling themselves around the atom, and up and down copper wires, in circular sorts of ways. They are fundamental to Quantum Theory, because Quantum Theory depends on the idea that sub-atomic particles can be treated like waves, and waves (as anyone knows who's played at the seaside) tumble you about in circles. For any subject in which the notion of back-and-forth or round-and-round crops up, imaginary numbers are close behind, panting to get a bite of the fun. They could be used to describe barn dancing if you like, because that involves the occasional swivel, and the twirl of delightful skirts.

Imaginary numbers are often represented by the letter i.

At Eton, Simon's mathematics left this practical stuff behind. It entered the world of magic.

---

won, and so did the woman in the shop. But when she scanned the card the machine said I hadn't.

"I phoned Camelot and they fobbed me off with some story that −6 is higher—not lower—than −8 but I'm not having it."

Scholarship report on ...... *Norton 8P* *12 3*

with the compliments of the Master in College.

Latin Grammar	*good*
Greek Grammar	*very good*
Latin Verses	*very good*
Mathematics I	*} !!*
Mathematics II	
French	*good*
Essay	*nothing much*

Extract from Simon's entrance exam report for Eton. He was awarded the highest scholarship score in the history of the school.

Using chemical analysis, we can determine Simon's age at the time he produced this drawing:

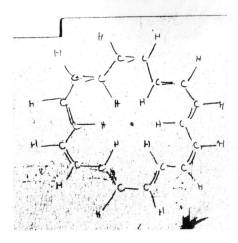

Is that a fly in the bottom right-hand corner?

It's a hexagonal hydrocarbon ring; it appears on the inside cover of a blue notebook I've discovered in a box file (yellow, mottled with mold) among the early geological layers of the back room of the Excavation. Hydrocarbon rings belong to O-level chemistry. I date Simon to be fourteen at the time he made this diagram. There follow two pages of trivial calculus of the sort designed to make schoolboys miserable all weekend; then come eight pages, in red pen, of boggledom.

They are not in Simon's hand.

They are in the louche strokes of Kenneth Spencer, his tutor: a list of bitterly difficult books and theorems by authors whose names look as though they've been put together in the forests of Eastern Europe from reject scraps of algebra: Sierpinski, Kuratowski, Tarski, Gödel, Banach, Schröder-Bernstein, Heine-Borel-Lebesgue, Zorn.

Spencer had been a senior mathematical scholar at Oxford and a university lecturer.

It's not the complexity of the problems that make his suggestions difficult, it's their simplicity. As with the difficulty of finding a way into Groups, these problems about the foundations of mathematical thought are as hard to clutch hold of as an oily ball; they are greasy with simplicity. The things seem hermetically sealed against all forms of assault.

When I was Simon's age (fourteen, at the time of the blue notebook) I crept off to a friend's house to watch *Star Wars*, and see Luke Skywalker fly his X-Plane down at the Death Star. Skywalker—battered and bumped by Darth Vader's proton-beam ack-ack guns—finds a bombing run, a furrow in the impenetrable surface; roils down this valley between mountainous blocks of spaceship windows, the air boiling with enemy explosions; and drops his bomb . . . ever so neatly, a moment of supreme cinematic silence . . . into a tiny ventilation outlet (the only known architectural weakness in

the evil Vader's vast edifice); then roars up and away as the
Death Star erupts.

Eton beak and housemaster who helped and inspired pupils to excel in Maths

... A fervent supporter of the mathematical tutorial system,
Spencer taught several generations of top maths specialists,
helping to secure many scholarships to Oxford and
Cambridge. Notable amongst his pupils was SP Norton, a
King's Scholar whom other masters found almost
impossible to teach in a class since he was so far ahead of
everyone else.

Spencer gave up considerable spare time to teach Norton
privately, as a result of which Norton (now an independent
researcher at Jesus College, Cambridge) obtained a First
Class external degree from London University while still at
Eton. He used to play three-dimensional chess with Norton
and another mathematical scholar, Campbell. The two boys
were better at this complicated game than their beak, but
often devoted so much energy to beating each other that
Ken Spencer would win ...

An obituary is supposed to be about the person who has died. Kenneth
Spencer's obituary in the *Daily Telegraph* spends much of its length
talking about Simon.

The mathematics that Simon was investigating was like this
also: it required unexpectedness, cavalier courage, total
precision, utter clarity of thought, and a mountain of
insolence.

On a table beside the moldy box file containing the blue
notebook there's a black-and-white picture of Simon, aged

fifteen: a thin, handsome chap with short, dark hair and an angular face. His jacket is too large; it drapes from his shoulders. He's receiving an award from Prince Philip, the Duke of Edinburgh, for winning a national mathematics competition. Philip is leaning forward with an air of paternal concern: "Is this little fellow right in the head?" his look says. "Why does he seem so beaten-in?"

He is tired, Duke—a little boy, fresh from wars.

After Spencer's eight pages of red ink, carefully laid out, grouped into sections—the whole thing headed by the word "Analysis"—come three pages of Simon's tongue-biting script, in blue ink. He is growing up fast: this is now recognizably the handwriting that Simon has stuck with into adulthood. It's a little broader, there's still a bit of bounce about the way the letters stretch across the page; he's not yet gnawing his tongue quite as hard as he does today.

$$= (bd, by)$$
$$\text{nw} + \text{n} = (w+n)a$$
$$1a + \text{wn} = a(w+n) = (w + na)\,a.$$
$$c) = (r, d, g)$$

$$rw + c = (dw + e)(rw + g) + \text{ww} + h.$$
$$c = \quad e + h \qquad\qquad \textit{premodular}$$

From the blue notebook: numbers (except a few token ones floating by) have almost disappeared.

Something takes place during these three pages after Spencer has made his recommendations for further reading. There's a shift in Simon's attitude to mathematics. A mathematician would no doubt know immediately what's going on with the marks he makes, but to a non-mathematician there's an unsettling calm. The clunky equipment of school methods (columns of long division, times-tables, logarithms, cosine rules) has been pushed aside. It's all letters. It's as though Simon's found a way under the surface cladding of mathematics into its levers and cogs.

Then we are back to Spencer's red pen. There's been a change here too. Spencer is no longer at ease. He has lost his patrician cool. Instead of the pages being blocked out, divided into sections relating to different theoretical points, references to higher authorities appended as appropriate, he's dashing at the page, scrawling wildly:

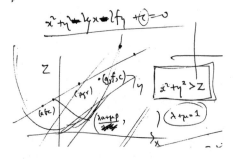

"Differential Geometry." Spencer batters on:

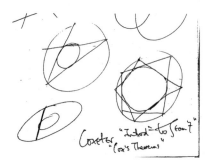

Spilling knowledge as fast as he can now, Spencer's handing out every last weapon to Simon that he can think of, before the boy disappears so far inside the soul of the subject that he can no longer reach back to ordinary men:

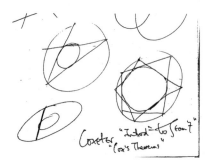

Then, ***pooof!***

Spencer vanishes. His hand is never seen in the blue notebook again. A few pages later, that disturbing shape from the back of the front cover reappears. It's simplified—still chemistry but pared back:

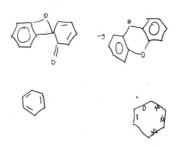

Simon is doing to this hydrocarbon what he has just learned to do with numbers—getting under the surface of it, looking for the machinery, the general pattern, the cogs and levers that make it function. Over the next pages, the last traces of oxygen vanish. Released from particulars, the purified form starts to flex:

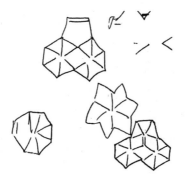

mutate:

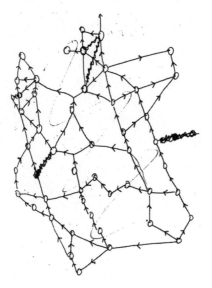

(Oops, no, sorry, that's a set of
train/bus routes from Cambridge,
Bedford and Huntingdon that Simon's
trying to work out on the same pages)

and tessellate:

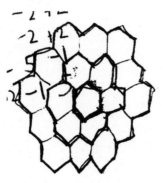

as Simon drags it with him into the depths:

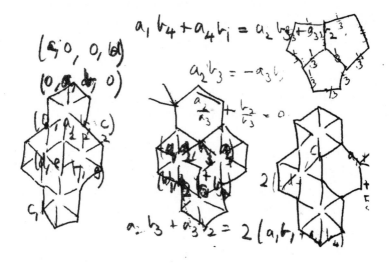

October 1966. Eton allowed him to start a university degree in pure mathematics. Simon calls it "My Day of Awakening." His "Arrival." "The moment I felt I really came out." But he is not talking about mathematics.

He remembers the number of the bus he took to college on the first day: the 441, which was "a perfect square: 441 = 21 x 21." ("No, Alex, that was the bus that took me to Holloway, not Imperial. The bus to Imperial was the 705.")

He remembers the locker room, the woman at the desk who gave him the key for his own locker, where he could store his books on topology, algebraic geometry, complex numbers, infinite series, Set Theory, Group Theory; the locker door— metal, gray, holes at the top.

("Alex, I do not remember holes.")

He remembers opening the door. He can picture the first sight he had of the locker's contents, inexplicably left behind by the previous owner.

And after that he forgets. The rest of his three years at Imperial College and later Royal Holloway College are gone.

It was what was in the locker that changed Simon's life that day.

London Transport folders.

To this biographer, the only way to make sense of the moment is to fill it with coruscating splendor: piled from bottom to top, the metal cupboard tumbled with inch-thick custard-yellow timetables: Merry Maker trips across the aching moorlands of Yorkshire, rambling tours of Cornwall's sea-blathered cliffs, Weekend Specials to the ghost-filled churches in Lanarkshire, Magical Mystery Tours of the elfin forests of Northumberland. Stashed on the locker's shoe shelf: fat towers of unused shilling-return bus seats held together by elastic bands; dangling from the coathook, an epiphany of Ordnance Survey Trailfinder maps crammed into Woolworths bags.

Simon insists it was nothing of the sort.

Ask him directly what he found that day and he gives a Grunt Number One (the Mother/Loveliness Grunt), swelling momentarily to an unexpected Number Two (Father/Puzzled) and concluding, *allegro,* with a staccato riff on Number Four (Frustrated).

"I'm sorry, but when I came across the London Transport folders I had no idea that I would be asked to remember every detail of the incident forty-four years later."

To get fuller details, you've got to write to eSimon.

Explanation for Grunt #1: "As I say, it was when I was taking my degree that I really felt I came out in terms of public transport . . . What you can say is that I did NOT have a sudden epiphany. I remember there were exactly six different types of folder in the locker. I kept one of each, and gave the rest to the office which had issued me with the key to the locker."

Explanation for Grunt #2: "They were free publicity. But I didn't know that until I had looked through it and was confused. I decided to follow up the sources of information mentioned in the folders (e.g., go to London Transport's publicity office and buy area timetables). It was this process that initiated my lifelong fascination with public transport."

Explanation for Grunt #4: "I wish that every fourteen-year-old had had a chance to discover the joys of buses the way I did—if so, this country wouldn't be in the mess it is, and incidentally . . ."

"Yes?"

"Why have you not included a Grunt Number Three?"

That's all Simon has to say about his time as a London schoolboy undergraduate. With the close of this single, dazzling memory, London University, Simon's alma mater, is gone.

S.T. 3 Aug 69.

Eton boy wins first class degree

By Tim Devlin

A FEW WEEKS after leaving Eton at the age of 17, Simon Norton has won a first-class honours degree in mathematics at London University.

He does not look like a swot and he remains modest about his achievements. He leans over backwards not to boast but at the same time to be accurate. When I suggested to him that he must be one of the most brilliant mathematicians of his age in the world, he said: "I should hope so."

His recreations are chess and bridge, which he began when he was aged about five. His ambition? "I want to break new ground not necessarily by inventing something, but if I had my dream it would be to leave the same kind of mark on history as Einstein did."

Sunday Times, August 3, 1969.

"Never heard of it."

"But you used to play it, Simon! All the time, at Eton. Three-dimensional chess. It says so here in Kenneth Spencer's obituary."

181

We were in my study; Simon was working through his "backlog" of post. He'd been in Canada at a Monster conference for two weeks, and in that time several hundred letters, some from the Office of the Deputy Prime Minister, had piled into the house. To his left, beside his knee, was the pile of unopened letters. He had thrown them down in preparation, like the toy bricks he used to play with as a baby, and was now picking them back up, item by item, and investigating each with forensic delicacy—its lettering, stamps, emblems (if from the government or royalty), postmarks. He was sniffing for interest, although that must be the wrong word, because Simon has no sense of smell; but this dog-like detective work appeared to be led by his nose. If the packet he was analyzing at that moment had been human, it would have punched him on the conk for a perv.

I snatched up the photocopy and shook it at Simon's nose. "Here, paragraph four: 'He used to play three-dimensional chess with Norton and another mathematical scholar, Campbell . . .'"

"I don't know what he's talking about, hnnnn."

He took the sheet, pressed it still closer to his eyes, and studied the print as though the letters weren't to be trusted even now. "Written by a journalist. Someone in your line of work," he said maliciously.

"Nonsense. Obituaries are written by people who knew the person, especially obituaries of Eton types." I still felt rather bitter about the fact that when I'd sent my obituary of my last biographical subject, Stuart, to the *Telegraph,* they'd wanted to change the words round to portray him as a grateful, reconditioned savage, and cut my favorite comments (on grounds of "house style," if you please; then, when I protested, on the grounds that they were rotten sentences). But the *Telegraph* does not make up long-winded, independently verifiable facts about Eton beaks. If the obituary says Simon used to play three-dimensional chess with his tutor, Simon played three-dimensional chess with his tutor.

"But what," I pressed, "is three-dimensional chess? Do you play it in a cube?"

"I have no idea," retorted Simon, becoming flustered. "I don't want to think about it just now."

"I mean, if you play two-dimensional chess in three-dimensional space, does that mean three-dimensional chess has to be played in four-dimensional space?"

"Aaaah, hhnnnn . . . uuugh . . ."

"There's a mention here on the Internet that it's played in *Star Trek*. They seem to be using one of those cake stands you find in village tea shops. Was yours played on a cake stand?"

"Hnnnn," he gave a Grunt Number Four (frustrated) and eyed a savaged Jiffy bag. He dropped the wreckage onto the pile next to his right knee, ready to be gathered into the green wheeliebin, picked up an envelope from a thermal underwear company and assaulted the plastic window.

"Please be quiet. I have to concentrate now."

Always, in these disembowelings, he takes care to include a windowectomy, and set that non-recyclable material aside, ready for transport to his black wheeliebin.

Goodness gracious, where was Simon when Simon was about? Here it is, in the moldy yellow Eton box file I took from the back room of the Excavation: the rules and board for three-dimensional chess. Simon not only played the game, he invented it.

*21 3D chess

It's *NOT* 3D chess, Alex, please try to get this right: it's three-*player* chess.

Simon

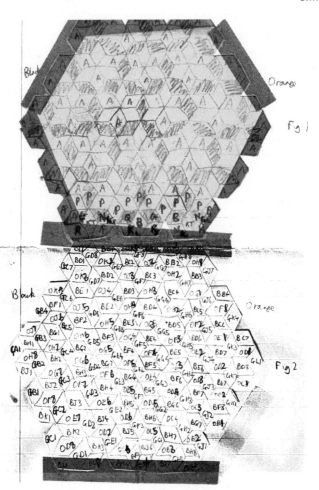

3 Player Chess

(as invented by Simon)

1. Pieces laid out in Fig 1. Squares in Fig 2. Green has his pieces at the bottom; his left-hand opponent is Black, his right-hand opponent Orange.

2. Each person has two kings. One, the left king, may be checked only by the left-hand opponent, and the other, the right king, only by the right-hand opponent. The kings are marked with a cross according to the color of the opponent who is allowed to check it.

3. There are nine kinds of men:
   a. 2 kings, cuboids with crosses, marked as above (K)
   b. 1 queen, a cuboid with a circle on top (Q)
   c. 1 general, a cuboid with a triangle on top (G)
   d. 2 rooks, cylinders (R)
   e. 3 bishops, cubes (B)
   f. 2 ordinary knights, paper clips in cardboard bases (KT)
   g. 2 gunner knights, triangular prisms (faces missing) with circles on top (Ng)
   h. 2 anti-gunner knights, like gunner knights except for dots on top (Nag)
   i. 13 pawns, tetrahedral

4. Captures are made as in chess, by moving onto a square occupied by the piece. The pawn is an exception that will be dealt with later.

5. The moves of the pieces are as follows:
   a. The bishop moves parallel to the edges of its rhombus, so as to stay on squares of a certain color, e.g., B(GG5)—DE3.
   b. The rook moves perpendicular to the edges of its rhombus, e.g., R(GH6)—BG3.
   c. The queen moves like the rook or bishop.
   d. The king* moves like the queen, except only one rhombus at a time.
   e. The gunner knight moves one rhombus B-wise and one rhombus R-wise, to make a move that cannot be done by the Q e.g. Ng(OG6)—GF6.
   f. The anti-gunner knight moves like the gunner knight, reversed.
   g. The ordinary knight moves like one of the two preceding knights.
   h. The general either moves like an ordinary knight or moves along its long diagonals, e.g., G(BD6)—BD2.

6. Pawns, when on rhombi pointing to their possessor, move directly forward one rhombus. Take R-wise left or right forward. When on other rhombi, they move directly (R-wise) forward one, two rhombi, or three first time. They take R-wise or B-wise left or right forward. If they move two, three rhombi, they may be taken by a pawn as if it had moved only one rhombus. All pawns may be promoted to any piece (except K), on reaching rhombi from which they cannot move. E.g., (for green ones) P(GF3)—GF4, P(0D2)—BH4, P(BF3)—BE2, P(GE5)xQ(BE3), P(0G4)xR(GE5), P(BD5)xK(GG6), P-GE7 = Q.

---

*It is possible to exchange the two kings if both are in check.

The game is based on that hydrocarbon shape Simon purified in his notebook. It has re-emerged, bubbled back up to the surface from his mathematical depths, confabulated, fluted, overlaid with letters and recaptured numbers, Sellotaped to a sheet of lined file paper, and ready for use in your tutor's pipe-smoke-filled room.

The top half of the illustration on page 184 is the playboard. As an ordinary chessboard has black squares and white squares, a three-dimensional chessboard (for three players) has black diamonds, white diamonds, and diamonds labeled "A."

The bottom half gives the location reference for each diamond.

Here's a tidier version of the playboard, with the diamonds labeled "A" replaced by gray-shaded diamonds:

Each of the three players has:

Two Kings (K) 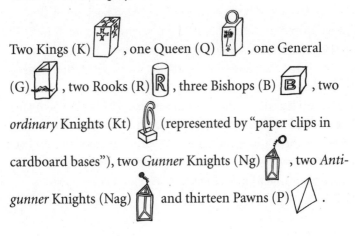 , one Queen (Q) , one General (G) , two Rooks (R) , three Bishops (B) , two *ordinary* Knights (Kt) (represented by "paper clips in cardboard bases"), two *Gunner* Knights (Ng) , two *Anti-gunner* Knights (Nag) and thirteen Pawns (P) .

*22 Breakthrough

Hurry! Hurry! Back to the classroom! Time for more misdeeds with Triangle and Square!

To recap: Square and Triangle are symmetrical. You can do all manner of things to them—rotate them, flip them over, in any combination you fancy—and they'll *always* look just the same after your wickedness as before:

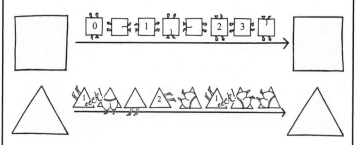

In short, Group Theory is the study of the moves that *always* leave an equilateral triangle and a square (or other symmetrical object) looking the same. To show how these different moves combine (e.g., one turn of a square, followed by one more, equals two turns in total), we use a sudoku-type grid called a Group Table . . . Oi! You, yawning at the back! Pay attention!

How does a subject that's for children—cutting squares and triangles out of dayglo paper, representing them being bullied and shoved about by symbols such as   and   — lead to an intellectual discipline so intricate and metaphysically profound that it can absorb a genius like Simon for the majority of his adult life? These Group Tables for the turns and twists of Square and Triangle seem too bland for discussion. Their straightforwardness as secretarial devices makes them appear beyond analysis, stagnant with simplicity. How could there possibly be anything extra to say about them?

But there is.

The mathematics libraries of the world are crammed to the ventilation ducts with books that depend on an easy idea, gently hidden inside the Group Table for the rotations of Square but missing from the Table for Triangle.

It's called a "Subgroup."

Square's Group Table of turns has one, and Triangle's doesn't.

(There's a slight fuss about what Simon calls "trivial" Subgroups, which all groups possess, but we'll ignore this complaint here.)

A Subgroup is simply a Group within a Group, a smaller symmetry hidden inside the larger one. The Group Table for the turns of Triangle doesn't contain any Subgroups. That's why it's considered an "atom" of symmetry. It can't be broken down into smaller symmetries.

A square's Group Table of rotations does contain a Subgroup. It can be further broken down in symmetry terms so, therefore, it is not an "atom" of symmetry but a flibbertigibbet.

Subgroups are one of the keys to turning Groups from a distraction for children into a mathematical smash-and-grab raid for Simon Nortons.

In Chapter 25, we'll see how to spot the hidden Subgroup in the Group Table of Square.

A last paragraph, for advanced readers: the "atoms" of symmetry are to Group Theory what the prime numbers are to whole numbers—the building blocks of the entire system. Just as any whole, positive number can be constructed by multiplying together prime numbers, any finite Group can be made up by combining together these "atoms" of symmetry. For example, the number 15 is composed by multiplying together the two primes 3 and 5. Similarly, the rotations of a pentadecagon (a fifteen-sided regular shape) can be constructed by combining the rotations of a triangle and the rotations of a pentagon.

23 Breakdown!

I don't feel that the episode of the toilet falling through the floor has anything to do with my character.

Simon

PING! An email, from Simon:

Friday, October 22nd, six minutes before midnight
Subject: detailed comments on your book
Let me start by explaining about Pasquino and Nick. You don't know about Pasquino. He was an Ashdown boy. At one stage he and someone called Carpenter were my best friends there—well it might be more accurate to say my only friends. Then one day they joined the herd of bullies. I may be able to forgive Malcolm who, after all, was bullied himself, but I will never be able to forget Pasquino and Carpenter.

Nick is another traitor, this time not just against me but against most of the country. You do know Nick—Nick Clegg. Thanks to the Cameron Cuts he's signed up for, many of the bus and train trips I've been doing recently, and even more that I haven't got round to doing, may no longer be possible.

Now let me remind you of your promise that I would be allowed to vet all material for publication. I accept that I cannot have a 100% veto but the general tenor of what you have written is so far from acceptable that if you go ahead with it in anything like its present form you will have broken every one of the promises you made and will thoroughly deserve the description of traitor.

It was with difficulty that you managed to persuade me to include a description of my rooms; you said that without that any biography would be incomplete. I reluctantly agreed on the premise that you would do your best to tone it down and that you would help me tidy it up so that by the time the book appeared I would have nothing to fear from any possible visit by housing inspectors.

Furthermore you said that I could use the book as a soapbox for the issues on which I care deeply.

So what has happened? Far from toning it down, you have done your best to emphasise the issue by putting the words which you say I won't allow you to use, but which an astute reader will be able to deduce, in display format. You return to this again and again as if to make sure that even the dimmest reader is aware of how awful my rooms are. You use the insulting word "excavation" repeatedly. You explicitly mention my fear of the housing inspectors thus ensuring that any of them who read the book will know exactly where to go to find something wrong; and as your interpretation of helping me to tidy up consists of doing perhaps 0.1% of the work and expecting me to do the rest all alone, there's little chance that I will be able to rectify things before they arrive.

And to cap it all, you confine all mention of my campaigning activities to the barest minimum, in spite of my repeated statements that they are essential for understanding my life (e.g. they account for my loss of interest in maths)! Not so much concern with completeness now, eh?

Thinking about my life I can detect four main turning points.

1966: Start work at London University: mentioned
Leave confining atmosphere of Eton: understated
Find cache of London Transport publicity: mentioned but with facts completely wrong.

1969: Start work at Cambridge: mentioned
Meet Professor Conway: mentioned

1985: Lose Professor Conway to Princeton: mentioned but understated
Bus cuts culminate in chaos of deregulation: ignored, although this is behind both my move towards public transport campaigning and the untidiness of my rooms (I couldn't keep track of all the timetable changes and eventually gave up trying).

2002: Mother dies leading to my purchase of a flat in London: ignored
I start to feel my age: ignored.

So you have either ignored or understated the majority of the key events in my life.

As I'm sure I've said, it was NOT going on trips that led me to lose interest in maths, it was getting involved in campaigning after the deregulation of bus services. Conway moving to Princeton also deprived me of the stimulus behind much of the maths I did.

I don't think there's any hope that your next draft will be fully acceptable, but I do think that if you take account of everything I say then only minor changes will be needed.

Now let me turn to the subject of truth. I think I told you how my enjoyment of one of Victor Canning's books, *The Finger of Saturn,* was completely spoilt when he referred to a railway viaduct as "long disused." It was on the Beeching hitlist but it survived and is today marketed as one of the major scenic routes of Devon and Cornwall. The idea that it could have been closed depressed me profoundly and spoilt my enjoyment of the book. So, while I accept the concept of writer's licence, I see no reason for many of the factual errors you have included, such as confusing Sharyn McCrumb's *Bimbos of the Death Sun* with her *If I'd Killed Him When I Met Him.*

Of course, if you prefer to get things wrong deliberately, as what you say on page 35 might suggest, you belong on the team of a trash publication like the *National Enquirer* as I said in my last message.

Now let me start on the detailed critique . . .

[There followed five pages of corrections.]

24

I used to worry at school that there was I enjoying
numbers when everyone else was made to do boring
things, like swimming or the American Civil War. Why
didn't the teacher tell them about the cyclic permutations of
142857?

Simon

"No, don't lie on the floor, please, Alex—do you mind if I call
you Alex?"

It was the hypnotherapist speaking. Andrew Cunningham: a
slight, soft-spoken man, the star of dozens of TV shows for
curing stage fright, fear of dogs and heights, and for cramming
the tremulous with braggadocio. For the Channel 4 program
Faking It, he transformed a city solicitor into a foul-mouthed,
shaven-haired garage music DJ.

I was going to be hypnotized. Hypnosis, I'd decided, would
help me investigate the curlicues of Simon's mind.

One Harley Street is the last chance a hypnotist has to work
in this prestigious part of London. It is the high-end numbers
of Harley Street that belong to the lucrative professions like
sucking out fat with a vacuum cleaner and burning women's
faces with acid. Fail to fit into some cubbyhole at Number 1,
and you drop off into Cavendish Square; a little farther south,
and before you know it your business is camping alongside
Clairvoyant Claire under Vauxhall Bridge.

I'd met Andrew in the entrance hall of this office block,
beside a brass plaque stretched thin with company names

suggesting empty one-room offices and unmanned telephones: Mindworks, Mindspa Clinics, Management Psychology Ltd, The London Therapy Center. We had to go up and down so many corridors and back staircases to get to Andrew's consulting room that possibly it was no longer in London at all.

"The first thing to know about hypnotism is that it doesn't exist," Andrew said with satisfaction. "Here we are, third door on the left." After the urn of potpourri.

My idea was that if you could identify what motivated a monomaniacal genius like Simon to do mathematics all day long, then ask a hypnotist to make you feel that motivation too, you won't be suddenly good at math, but you might begin to study it in the correct pioneering spirit and force yourself to be better.

The noise of traffic outside disturbed his curtains.

"Hypnotism can't add something you haven't already got," Andrew insisted. He's had success helping people to flush their cigarettes down the toilet because "Anyone can give up. The potential is already there. You're not making them do something they can't." Hypnosis "removes blocks to what is natural." It's similar to extreme daydreaming. "If you talk to people who've been hypnotized, they won't say, 'He clicked his fingers and I was under, then I woke up and I don't know what happened.' They'll say, 'I was aware, I just felt like pretending I was an artichoke.'" You can mesmerize people to eat lemons believing they're peaches, or to see green when they're staring at purple, and people really do see green. Clinical psychologists have done experiments. The part of the brain that sees green registers. Everyone's got the ability to be fooled.

Hypnotism is about "rapport: get the subject to feel responsive, then make suggestions that are easily taken up."

"Such as, 'Congratulations, you're a maths genius'?"

"Exactly, except in mathematics, you'll think you're a genius, but what comes out will be rubbish. Now, how precisely do you imagine that this is going to be useful?"

A good question—but I was prepared.

Mathematics, I've realized from Simon, is a branch of painting. You can go to fine-art classes, take a first-class degree, study the masters from caveman to Rolf Harris, practice musculature and composition until your thumbs split . . . and still be bad at painting: your faces, technically perfect, are dead. Your gestures, flawless from the anatomical point of view, chopped out of cardboard. Your compositions, correct according to every authority since the Greeks and the Golden Section, without suggestion. You have no feel for vigor, unexpectedness, life. You are brilliantly trained, and the best stroke you could make would be to dump your brush in the vegetable grinder. Ditto, mathematics: unless you have natural talent, everything you do will be insipid, mechanical and lack adventure. A hypnotist cannot suddenly create natural talent, but he can give one an obsession. If I could wake up in this peculiar Queen Anne armchair in half an hour's time, slavering to spot geometrical and numerical patterns, then I might also lie awake at night rabid with desire to characterize and make sense of these patterns. And *that,* I enthused to Andrew, grasping the air in my arms as though I were making off with an enormous coconut—the fruit of my delicious idea—is the essence of mathematical inspiration.

Andrew gave me a puzzled look, leaned to one side and made a mark on a pad of paper.

Mathematicians are aware if they're creative or not, I hurried on. When they discover that they are not, it shudders their soul. I have friends with PhDs in number theory who wince with embarrassment whenever someone calls them "a mathematician," because they know they are not worthy. They are to the authentic artist of mathematics what a weekend watercolorist is to Cézanne. At Trinity College in Cambridge, there was a man who scored a double first in his undergraduate degree, took his PhD in the flash of an eye, and still gave up in despair and became a tuba player. Since the age of twelve he'd doubted himself: under the sheets with a flashlight in his bedroom, he had not been able to pass a well-known childhood test of genius, i.e., to prove that the solution to any equation of the form

$$ax^2 + bx + c = 0$$

must be, for reasons known otherwise only to God and magicians,

$$x = \frac{-b \pm \sqrt{(b^2 - 4ac)}}{2a}$$

This man therefore believed—I lowered my arms at last, exhausted already by the algebra—that he could not *ever* be imaginative enough to call himself a proper mathematician.

Andrew tapped his fingers, beetled his brow—and made another mark on his pad of paper.

I'd read about how hypnotherapists work. Like biographers and mathematicians, they have to find the kinks in the surface of your problem before they can burrow into it. They pick on any catchy and summarizing phrases you tell them during the twenty-minute warm-up period, then repeat these over and over again in the sleepy bit to help drive you "under."

"Simplicity," I summarized. "What drives Simon is a search for simplicities. I asked him for a question to test whether somebody has mathematical talent. He proposed, 'Why is 11 x

11 equal to 121, and 111 x 111 = 12321 and 1111 x 1111 = 1234321?' Why, in other words, is that pretty rise and fall of the digits preserved?

"'It's not a question of *why*,' I'd snapped, 'they just do it.'"

But it *is* a question of why—and "why" is a question of discovering what is the simplest way to think about—visualize—sequences of 1s so that when you multiply them together you can see that *obviously* they must produce this charming answer with the bulge in the middle. I don't know the reason. I can't do it. 11 is just 11 to me, a pair of dumpy number 1s. I'd need to get out my pen and paper and devise ornate, well-trained formulas to parse 11 or 111 or 1111 into columns of units—tens, hundreds, etc.—then ram them together in a multiplicative frenzy that would leave x's and y's groaning all over the page. That would be my answer. But Simon does the opposite: he thonks down his pen; he sits back; he folds his hands behind his head; he makes one of his ghastly attempts at a whistling noise. The one thing all good mathematicians hate is honest labor. Instead of calculation, his mathematics requires an ability to see things in a different light, like those optical-illusion games in which you have to blur your eyes to turn a splatter of dots into a three-dimensional picture of an advancing stegosaurus.

"Right, Alexander! Feet flat on the floor. Relax and close your eyes. *Ten* . . . math is about finding simplicities . . . patterns . . ." Andrew begins in a church-crypt murmur. ". . . *nine* . . . don't divide up rows of ones and ram them together . . ."

That's when the trouble starts: I can't close my eyes. They flicker like beetle wings. Every inch of me starts to itch; every object on the street bellows. That isn't a van honking, it's a barricade of French juggernauts. That's not a scratch in my ear, it's a weevil boring toward my eye. Instead of feeling sleepy, I've jumped in the opposite direction: I'm less hypnotized than ever.

"... *eight* ... simplicities are everywhere and simple and you want to find them," counts Andrew.

A jackhammer starts up. Ants are making a nest in my nose. Most of us experience mild hypnotic states every day. Tests have shown that people blink less when they're shopping—they've entered a state of mild hypnosis.

"... *seven* ... simplicity is joyful ..."

But not for me. Andrew could go on counting down until he squeezed out the other side of zero and climbed to minus 1,000. I've never realized until now how infuriating the world is. Traffic noises fly through the window and bounce about the room like fifteen grasshoppers. Balloons filled with babies' voices burst against the glass.

"... *six* ... patterns such as 'two to the power of twenty-three spells sissors' ... *five* ..."

I try imagining the street full of integral signs climbing drainpipes, number sevens strap-hanging in buses, prams carrying Platonic solids—anything to get me into the world of math and escape this hyper-wakefulness.

"... *four* ... next time you see a car number plate you will want to factorize it ... *three* ..."

Hypnotizability has nothing to do with intelligence. Roughly 10 percent of people are good hypnosis subjects. The Nobel Prize winner Richard Feynman describes a hypnotist sitting him down and saying, "You can't open your eyes," and Feynman couldn't:

> I said to myself, "I bet I *could* open my eyes, but I don't want to disturb the situation: Let's see how much further it goes."
>
> It was interesting: you're pretty sure you could open your eyes. But, of course, you're not opening your eyes, so in a sense you can't do it.
>
> Richard Feynman, *Surely You're Joking, Mr. Feynman!* (1985)

"... *two* ... your eyes are very heavy, so heavy ... *one* ... so very heavy, it is impossible now for you to open them ..." Andrew concludes in a soporific clincher.

I open my eyes, giggling with embarrassment for both of us.

One useful approach with therapeutic hypnosis is to find a structure, already present in the client's mind, then doll it up with fresh encouragement and new suggestions.

Not long ago, a ten-year-old stepped into Andrew's office—extraordinarily, he had exactly the same grouse as mine.

"I want to do sums better."

It's not such a peculiar thing to ask for.

"What are you good at already?" Andrew asked. "And why?"

"Stories. Words come to me through a door."

"What color is the door?"

"Red. With a big gold handle."

So Andrew put him in a trance, took him back to the door and started asking mathematical questions: "What's two times two?" A four sauntered out.

"It was blue," remembers Andrew.

"Three times three?"

Nine made an entrance, dressed in green.

"Two times four?"

Eight stepped forth. Pink.

"And this was not a weird child," said Andrew. "Just an ordinary young boy." Numerals had been hiding on the other side of the door all along. Hypnosis simply made them less bashful.

"I, um, ah," I said, slapping my palms together to give me courage when I got back from hypnosis, "have been studying Cayley's mathematics myself a little, as it happens."

"Hnnn," grunted Simon.

We were walking in Cambridge along the Backs—the section of the river that runs behind the ancient colleges—and turned off it up the path leading past the library, on to the outcast newer colleges and university cricket pitch.

"I gather that though Cayley published papers of great length, many are now considered somewhat arcane," I said.

"Aaaagh," from Simon.

"It's his shorter notes that are thought to have captured the deepest insights. A vindication of his belief that the best work of a mathematician, like that of a biographer or artist, was often that done in five minutes."

"As I say, I don't know that he did say that."

"I know you don't know. That's why I'm telling you. He did."

"If you say so."

"I do say so."

"Hnnnn."

"For example," I kept going, "the idea in Group Theory that is named after him: Cayley's Theorem. That was one of those flash-of-insight triumphs."

"What's Cayley's Theorem?"

"What's Cayley's . . .! Simon, you haven't lost your marbles to that extent! Cayley's Theorem: a fundamental theorem of Group Theory."

"Never heard of it."

"Well, let's see, how does it go?" I gathered in a pedagogic breath. "Yes, I think that would do it . . ." I added in a tone of subtle calculation. In fact, I had spent several hours memorizing it that morning from *Schaum's Outlines* ("The perfect aid for better grades!") "Let S be a semigroup with Identity . . ."

For those of you not familiar with this language, note my use of the term "Identity." Since hypnosis, I have had insight into Identity. Identity is much more important to Groups than it is to humans. Humans can have multiple identities, fractured identities, confused identities; identities that they've accidentally put in the dustbin and someone has stolen; identities that have wandered off to Thailand and for which the owner has to take six months' sick leave to rush after and find. With Groups the situation is simple: a Group must have a very clearly defined object called an Identity Operation, or it is not a Group. Any Group Identity Operation claiming to be complex, split or misunderstood is simply a liar. It should be instantly locked up.

For Groups an Identity Operation is one that does precisely bugger-all. It's a scandal even to dignify it with such an energetic word as "Operation."

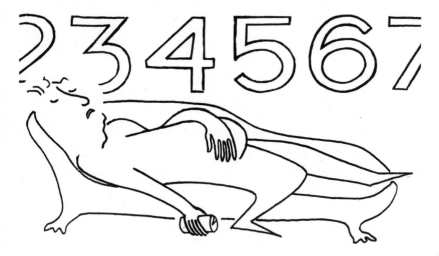

The Identity Operation.

An Identity Operation leaves an object in exactly the same state as before. That's how lazy it is. With Square, you can do various things that will leave a square *looking* the same as before; but you know, secretly, that these things have involved furtive twists and flips and turns. There's only one operation that actually represents doing nothing to a square:

That's the Identity Operation. With Triangle, the comparable Identity action is:

The Identity is everywhere in mathematics. With ordinary addition of numbers, what do you have to add to leave the number exactly as it is? Zero:

$$7 + 0 = 7$$

"Adding zero" is the Identity Operation for addition. For subtraction, just as for addition, 0 is the Identity:

$$7 - 0 = 0$$

7 minus 0 changes nothing; 7 is still 7.

This doesn't make zero the Identity in all calculations. In multiplication, zero kills everything off:

$$7 \times 0 = 0$$

It's the number of annihilation in this case, not Identity. Everything zero touches in multiplication, it lays to desolation:

$$3 \times 0 = 0$$
$$43{,}857 \times 0 = 0$$
$$85{,}417{,}023{,}207 \times 0 = 0$$

So, if zero *isn't* the Identity Operation for multiplication, what is? What number can multiply with every other number without ever changing any of them, even slightly? The number one:

$$7 \times 1 = 7$$
7 x 1 changes nothing: 7 is still 7.

Context in Identity Operations is everything, just as it is in life. Just across the park Simon and I were walking through is Charles Darwin's old house. As a young man, all Darwin had wanted to do was race dogs, catch rats and get drunk. But, as with mathematics, so with humans: circumstances make all the difference. When Darwin gave up the sauce and pushed off to the Galapagos Islands on the *Beagle,* lo! that wastrel became the volcanic intellect of the age.

Any number can be an Identity if you make the circumstances right. Take twelve. It can't oust zero as the Identity number for ordinary addition. It's much too energetic:

$$7 + 12 = 19$$

In subtraction, twelve soon results in negativity and gloom:

$$7 - 12 = -5$$

With division, twelve is too persnickety:

$$7 \div 12 = 0.5833\ldots$$

For multiplication, it loses self-control. It sends numbers racing into the distance:

$$7 \times 12 = 84$$

But every day, without thinking, billions of people around the world use a different sort of arithmetic in which twelve is the Identity: with clocks. Add twelve to any number on a twelve-hour clockface and the difference is bugger-all:

$$3:00 + 12 = 3:00$$
$$7:45 + 12 = 7:45$$
$$5:30 + 12 = 5:30$$

Now, to get back to my explanation of Cayley's Theorem to Simon, as we walked out of the park and along the Backs toward the Mathematics Faculty.

"Let S be a semigroup with Identity . . ." I repeated manfully.

It doesn't matter what a "semigroup" is: I'm using it here just because it sounds good.

"Let S be a semigroup with Identity; then by Cayley's Theorem there is a monomorphism of S into M_S, where M_S is the semigroup of all mappings of the set S into itself."

We reached a street corner, stepped out to cross, and Simon paused, apparently confused. The passing cars did not honk. They swerved gently to avoid us. Donnish behavior is well understood at this road junction in Cambridge. Then he sighed, shook his head to release the furrows from his brow, and we hurried over the tarmac to Selwyn College, kicking at the fallen cherry blossom.

"That's so elementary," he said, "I didn't even know it was a theorem."

*25 How to bag a Subgroup

Oh dear, oh dear, here he goes again.

Simon

The way the turns of Triangle combine can be presented as a table. This is a secretarial device for keeping track of all the possible results, and is read in the same way as a distance chart at the front of a road atlas:

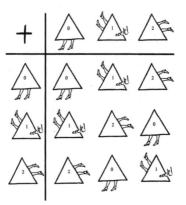

Similar Group Tables can be drawn up for the symmetry operations of Square, and any other symmetrical-looking shape—a rectangle, a pentagram, a cube, a tetrahedron . . .

What made these secretarial devices become an area of interest for mathematicians was a discovery made by a French schoolchild in the nineteenth century. He realized

---

that some Groups had hidden inside them smaller Groups—smaller symmetries, minding their own business, inside the larger one, like the cogs inside a watch.

The existence of Subgroups meant that certain Groups could be broken down into more basic units, just as molecules can be broken down into atoms.

The next question for mathematicians was, therefore, how many of these fundamental atoms of symmetry are there, and what are they?

A Subgroup is not different from a Group in any way. The prefix "Sub" just means it's a Group locked inside another Group. Sometimes Subgroups are easy to spot, other times they're as difficult to squint out as ghosts. But chop one away from the containing Group, and it's a perfectly good Group on its own. The "Sub" is a statement about context, nothing else.

Take the Group Table for the rotations of Square:

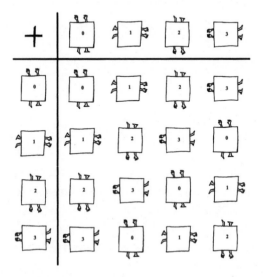

There's a Subgroup hidden in there.

To discover this Subgroup we need to understand a basic property of all Group Tables, called Closure.

The simplest way to think of this is as an aesthetic principle. One of the reasons Simon enjoys Group Theory is because this property of Closure is "beautiful."

For example, there are four possible symmetrical turns for Square that leave it looking exactly the same after the operation as before:

Leave Square   Rotate Square by
alone   a quarter of a turn

Rotate Square by   Rotate Square by
half a turn   three-quarters
                       of a turn

In the Group Table that keeps track of how these four operations combine, you never get anything else but one of these four figures with arms and legs on. Whatever symbols you put into the table are exactly the same as the symbols that come out. This is what Simon calls "closed." Nothing in the world you can do with these four operations is ever going to, say, result in Square being inverted, puffed up to double size, left teetering awkwardly on one corner, or shot through its belly:

Twist inside out   Fatten   Teeter   Shoot

A ridiculous example of a Group Table that is not closed is one where what comes out is something entirely different from what went in, for example:

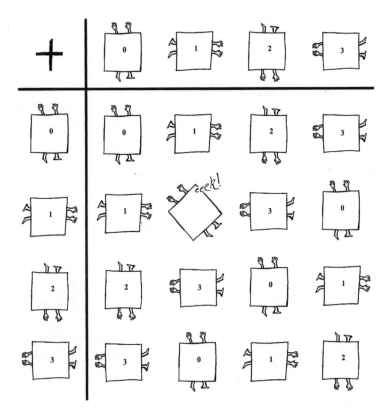

How did Teeter sneak in there? This has broken the Closure requirement (as well as just about every other requirement). Aside from the small fact that we already know

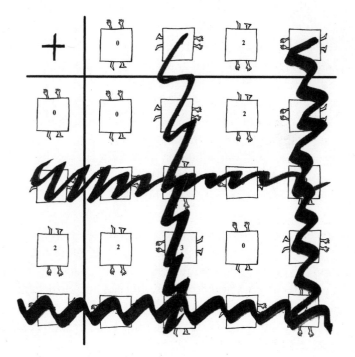

plus ▢ does not equal Teeter (and so something is profoundly wrong on that level too), Teeter was not one of our starting operations, so the table has an ugly, asymmetrical look about it. Some mathematicians might be interested in tables that have this type of thing going on, but not Simon.

Now, to return to the Group Table of Square. Let's eliminate two of the rows and their corresponding columns. Forget, for the moment, what entitles or motivates us to do this; just take a marker pen and get on with it:

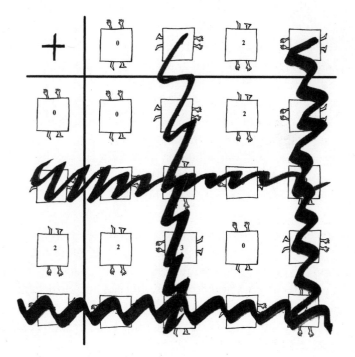

Now compress and tidy up the parts that aren't crossed out:

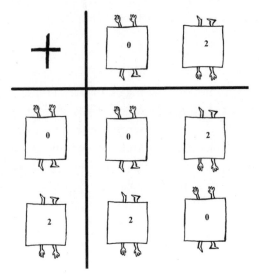

This is another aesthetically pleasing (at least in Simon's eyes) Table. It's closed. What goes in is what comes out. It's pretty enough to be a Group Table also, and since we know already that all the sums are correct, then it *is* a Group Table. It's called a Subgroup of the larger Group Table simply because we found it lurking inside the larger Group, but its balanced look and correct arithmetic mean it is a perfectly good standalone Group too.

Note: If we cut out rows in this high-handed fashion, we must also cut out the corresponding columns. What we are removing is the *entrance* of a certain symmetry operation into the Table, either through its row or its column. What displeases Simon (and the Closure requirement) is if, having removed all the possible ways a symmetry operation can get into a Table, its ugly mug *still* pops up inside.

We *could* have cut back the big Group Table in another way. Say, like this:

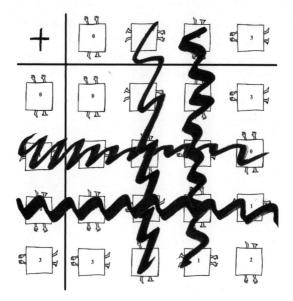

If we press together what's left over in this case, and tidy up, we get this result:

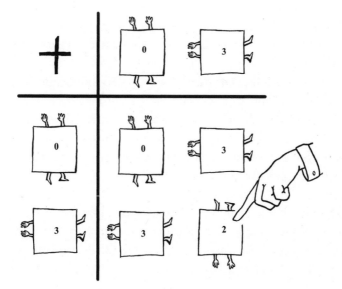

This again is correct as far as the arithmetic goes. It's been cut out of the larger Table, so it must be:

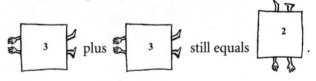

But as a table it's not aesthetically pleasing in Simon's

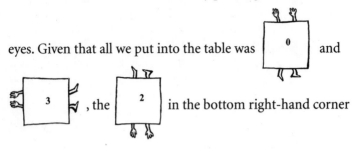

eyes. Given that all we put into the table was [0] and

[3], the [2] in the bottom right-hand corner

looks out of place and ugly and therefore contravenes Closure, the mathematical definition of a Group that's been specifically designed to prevent this sort of outrage. Group Tables should have an artistic appearance of balance, just like the operations they're describing, which is why Simon and so many other mathematicians are prepared to dedicate their lives to them.

This is a fundamental difference between the Group Table for the rotations of Square and the one for the rotations of Triangle: Square's Group Table has a Subgroup; Triangle's doesn't. As Group Tables get larger there are usually more and more Subgroups hidden away inside them, but not always. Some Groups preserve their purity.

The French schoolchild who discovered Subgroups was Évariste Galois, one of the great heroes of mathematics. Galois, born in 1811, nearly demented by genius and the failure of people to understand what he was talking about, once threw a blackboard eraser at his math teacher because the man was too slow; later escaped from his school by clambering over the playground wall; then marched through revolutionary Paris with a loaded rifle and was thrown in prison for waving a knife around at a party while making threatening remarks about the king. After being released, he was banged up again for some other absurdity. For several years he seemed on the brink of madness.

He died, "pierced through and through" in a gun duel over a barmaid. ("No, Alex, she was of higher status"—Simon.) The night before, he had written in the margin of his scribbled papers that revealed his astounding discoveries in Group Theory, "I have no time."

("In addition to a misquotation that is a romanticization.")
He was twenty-one.
("You have at least got that right.")

*   *   *

With the discovery of Subgroups, there's something constructive you can now do with a Group Table rather than simply use it as a tracking device for a triangle or square. The closest analogy I can think of is to a car—that's the equivalent of a Group. It's a nice object . . .

("No, it is not"—Simon.)

. . . does useful things . . .

("Such as destroy the world?")

. . . but if we smash this car apart . . .

("Good riddance.")

. . . you have two things: an engine (the Subgroup) and wreckage (all the other bits left over from the Group Table once you've extracted the Subgroup). Two things (engine and wreckage) instead of one thing (just the car) give you possibilities. You can now be creative about how you put them back together. Taking bits from the wreckage, you can bolt them onto the engine in new ways, to make a lawnmower, a microlight airplane, a go-cart . . .

("If you continue to talk about cars and their petrol-driven variants, I'm leaving the book.")

. . . or a giant egg whisk.

("That's more like it.")

Group Theorists spend a large portion of their time tinkering with Subgroups and their wreckage. The fellow with an old engine and boxes of oily bits at the bottom of the garden calls his workplace a shed. Simon calls his the Centre for Mathematical Sciences, Cambridge.

26

Simon has no oil on his feathers at all. He's one of the most completely honest people you'll encounter, and he takes it for granted that everyone is as honest as he is, so he's prey for every rogue painter and builder. He's the only landlord in Cambridge who dropped the rent when the poll tax came in.

Harriet Snape, former tenant

This book—my book—is changing Simon. New sounds have been massing under my floorboards.

Zwee*eeepppp*th-du**B**-thrrupp*PH*!—something rolled but abruptly stopped.

Trrr$$$$chchcx—plastic tearing.

Wh*eeee,* thun**K**! Wh*eeee,* thun**K**!—items flung about.

XxxccchuuuuhXXxxcchuuhXXXXcchuhxxcccchuuu
XXxxcchuuhXXXXcchuhhXX

Some of these noises have managed to squeeze through my carpet weave and let me know what's going on. "Wwwshshs-th*lump*-th*lump*-du**BUM**p": an object, very heavy, being dragged past the booby-trapped stairs. "K-p*shee*ow, KAH-***posh***, kluum*pff* ": rubble and wood shavings kicked aside. "Eee-urrgghg," the cautious opening of the Excavation front door.

Simon emerges with a rubbish bag.

Shocked by descriptions of the Excavation, he has started to clean up the g— (five letters)

"No!"

. . . the m— (six letters, archaic, think earth closet)

"No!"

. . . the s— (six letters, found at the bottom of lakes and coffee)

"No! *No!* NO!"

The *mess* that is carefully arranged into plastic bags in the Excavation.

"I am ashamed of it," he admits, dipping and shaking his head like a horse, clomping around the debris with a black dustbin bag. "Wait! What's that you just dropped in? Take it out! Oh dear, oh dear! King Alfred's *Winchester Bus Schedule, 1994.* No! No! I need to keep that. I refer to that often!"

By the end of this book it's likely I shall be writing about someone entirely different from the man with whom I began.

The reason he's ashamed of the mess is because "it's dysfunctional."

"What are you talking about? What's dysfunctional?"

"It."

"What's 'it'?"

"The mess."

"You mean the state of these rooms suggests that *you* are dysfunctional."

"No," he retorts, "*it* is dysfunctional. It affects how my life works. It functions badly."

The objects on the rut that leads to his bed are starting to compress into cardboard stone.

"What about if," I suggest, "instead of papers, it was rubble down here? No other inconvenience, no dysfunction, an occasional sneeze or two, because of the dust, that's all—would you mind that?"

"Yes. Sneezing is a functional issue."

"What about if it were vacuumed rubble? And a few carpets laid along the paths so you wouldn't hurt your feet?"

"Rubble that I can't move?" asks Simon, intrigued.

"If instead of plastic bags, it was rocks and pebbles, like at the seaside. If that table was a fat boulder. Those drawers filled with shingle. Everything just as it is, chaos and mayhem, but made of stone."

"I wouldn't mind that," says Simon, relaxing.

The whole process of cleaning up leaves Simon riven with anxiety: the refuse men might step through his patio ivy to complain that he's overfilled his wheeliebin so the lid pops open. They'll bang on the window, gesticulate at a colleague writhing on the pavement because of a slipped disk, city housing officials will demand words.

Because of the cleaning putsch there have been fresh discoveries:

Under an Asda Supermarket bag wave, two feet from his clothes cupboard, a cache of postcards from 1969. They are from teenage Romanian girls: Marieta, who signs her name, in brackets, like a hurried whisper; "Christina" ("I am at the sea-side and I send you a thought of friendship"); Lucia ("Maybe you will not know who is Lucia Mitrofan who writes you this view card"); Camelia ("When I saw your name, I hurried to write to you . . .").

Camelia's postcard has an efficient pencil tick (by Simon) in one corner. Why a tick? Because he's written back. A second card from Camelia has been found a yard away in a saucepan.

> Dear Simon,
> You asked me if I was in Czechoslovakia on August 20 and 21 [the anniversary of the invasion by the Soviet Union]. Yes I was. I think your question concerns the demonstrations of the people there. I didn't see myself the incidents, because in Karlovy-Vary all happened in the night when I was inside, but I feeled the tear gas in Prague.

Other finds include more newspaper clippings (collected by his mother) about Simon's startling boyhood triumphs in competitions (*Daily Sketch*). Glowing features in the *Times*

THE success of the British team at the schoolboys' mathematical Olympiad in Yugoslavia has been slightly overshadowed by the individual triumph of the team's youngest member, 15-year-old Simon Norton.

Simon, an Etonian, won the Gold Medal and a special prize for the most elegant and original solution.

Gold medal for Simon, the maths wizard

Exams

But he is still not through with his exams. When he returns on Friday, he

has to sit his A level exam in Maths and Chemistry.

"He is equally interested in chemistry and has not yet made up his mind what he wants to do. He's still very young," said his mother, Mrs. B. Norton, in St. John's Wood yesterday.

He scored 195 marks out of a possible 200 in the British Olympiad held to pick the team. The second boy in 10,000 entrants got 155.

His A levels should not prove insurmountable.

Educational Supplement, the *Daily Express,* the *Daily Telegraph,* the *Sun,* the *Star* (Johannesburg) and (his greatest supporting newspaper) the *Daily Mail* applaud his brilliance at the International Mathematics Olympiad.

The *Sunday Times* noted that while "the Russians and Hungarians had special coaching," the British team "were given no special preparation and seemed to treat the International Olympiad partly as a holiday." The French "explained away their dismal performance by suggesting that the questions were not 'avant garde' enough. 'They are the questions of Grandpapa,' said one of their officials." The only dramatic note "was struck two days before the competition ended. Professor Hanus Weinhart, from Potsdam, leader of the East German team, put on a rucksack and said he was going climbing in the mountains. He has not been seen since. The Jugoslav police say they believe he may have fled to Italy."

The clean-up has also turned up a heavy blue plastic wallet on a shelf in the back room of the Excavation, embossed to look like leather. Although covered in oily dust, the wallet smells fresh, like a child's seaside shovel; the shelf varnish cracked as I prised it off.

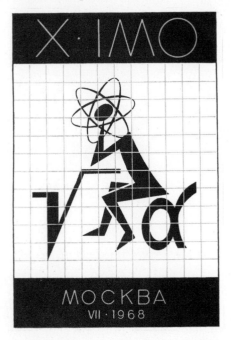

Inside, there's a nest of other folders made out of expensive textured paper. The logo on the largest shows a cartoon figure sitting on the letter α with, in place of a head, a ball surrounded by three electron paths. "Moscow VII•1968" reads the label: July 1968, a month before the Soviet invasion of Czechoslovakia. The midwinter of the Cold War.

Along the opening edge is a broken security seal.

Inside this, a folded sheet of paper with red writing. It has been made out to "Simon Noμton" (Simon Noshton) and is signed "Прокофвєв": P-r-o-k-o-v-ie-v.

НАГРАЖДЕННОГО ДИПЛОМОМ I СТЕПЕНИ НА X МЕЖДУНАРОДНОЙ

And if you have to ask what that means, you're too junior to know. Report to Prokoviev for execution.

The folder contains Simon's Olympiad answer sheets. He was allowed to keep them as a memento of the first victorious

D. MAIL. JULY 25 69

How Simon survives being a genius

BY JOHN MOOREHEAD

IF YOU happen to be one of those people like myself to whom the subject of mathematics i n d u c e s instant panic, then you would find it an unnerving experience to meet Simon Norton.

Simon is the 17-year-old Etonian who is regarded by his contemporaries a n d teachers as a mathematical genius.

He has just returned from Bucarest where he shared first prize in the 11th International Olympiad of Maths with a Hungarian and a Russian.

That's rather like a Russian bringing over a horse and winning the three-day event at Badminton. For Simon it completes a trio of first prizes at the Olympiads in consecutive years.

Mystery

How does he do it ?

That's something which is as much of a mystery to him as it is to the rest of the world.

'I don't know why I am so good at maths,' he says. 'I just got interested in them at the age of five and have been ever since.' * [see note, below]

Runs in the family, perhaps ? No luck there. Simon is the only mathematician among the Nortons.

His father is a St. John's Wood antique dealer and neither of his brothers has shown comparable academic brilliance.

The fact that he can't explain where his talent comes from merely increases, of course, the awe of the non-mathematician.

It's an impression which a brief encounter with Simon Norton tends to confirm. He's a tall, dark-haired, pleasant boy, much like any other of his age.

But when it comes to talking about himself, eloquence is not his strong point. It's almost as if he felt more at home w i t h figures and algebraic symbols than he does with words.

Did he ever, as a mathematician, feel isolated from his fellow human beings?

'That depends on what you mean,' he said. 'If you're asking whether mathematicians are isolated, then I think the answer is yes. But if the question is "Do they mind being isolated ?" then I think the answer is "No." At least I don't.'

Does he ever worry when people label him as a genius?

'No,' he said simply, 'I don't think it makes any difference to my life at all.'

PS. Here is one of the problems Simon had to solve in Bucarest :

Find an infinity of natural numbers as such that for any natural number n, $n^4 + a$ is not prime.

Answer : W h e n $a = 4k^4$, $n^4 + a = n^4 + 4k^4 = (n^2 + 2nk + 2k^2)$ $(n^2 - 2nk + 2k^2) = [(n+k)^2 + k^2] [(n-k)^2 + k^2]$. When $k \geq 2$, both factors ≥ 4 and hence $n^4 + 4k^4$ is composite.

PPS. Simon said this was one of the simpler ones. . . .

I've cleaned up all these articles. In life, they are orange-brown from sunlight, because his mother kept them displayed in a photograph album on top of her piano in London. They are toasted with boasting.

year, in which he scored 100 percent. What makes his work beautiful to read is again not its complexity but its simplicity: without drafts or false starts, he lays down his pellucid solutions to questions involving imaginary numbers, infinity and the distribution of primes, with the grace of a ballerina unfolding her hands.

Simon is starting to change me too, says my girlfriend, tartly.

I've taken to working in bed, in squalor; at fancy-dress events I appear as a number with arms and legs.

My doctor—I can hardly bear to admit this—insists that the pain in my right foot is g— I can't say the word. I have lost interest in my hair:

As Simon gets tidier to avoid me, I'm
becoming him. ("He," corrects Simon.
"Accusative case.")

*Answer to Fundamental Biographical Question Number 74,
subsection b, namely, Why write a book about Simon?* Because
he is to biography what the Monster is to the mathematics of
Group Theory: an intractable problem who nevertheless
represents an atomic type of being, a building block for
convoluted characters. I think Nature makes up our
personalities by mixing together a finite number of
fundamental different types—a pinch of adventurer, two
splashes of monomaniac, a dash and a half of Pompous Dad:
that would be a starting point for a banker. Simon is another
of these elemental types: the obsessive logician, untainted by
self-regard, sentimentality or any feel for romance. He's done
everything in his power to exclude the fact he's an emotional
human, except on the subject of buses and trains. Add a pinch
of Simon to the banker mix above and you might have a
health-insurance broker from Swindon, if you use too little; a

Dr. Strangelove madman, insanely compounding new breeds of atomic weapon, if too much; the most charming uncle, a champion bridge player in the mold of Omar Sharif, if just the right amount.

But how do you write a biography of a man who is pure Simon? It's like trying to write the biography of a hedge.

His face gets between me and my sleep.

Some days Simon loosens up.

"I once got a postcard from Julie Christie," he exclaims, popping his head round my study door.

"Really? Why? What did it say? When did you get it?"

"Huunh, uugh . . . I don't know. Excuse me, I need to fill out this voluntary questionnaire for landlords, from the City Council Housing Department."

Or, meeting me in the hallway as he's staring at the letterbox, waiting for the post:

"When I was on the bus from Cambridge to London, I was sick, but fortunately there was a sick bag on the bus, and I got rid of it at Trumpington, and the bus driver said, are you all right, and I said yes, and I was sick again. But that was when I got off."

"Thank you, Simon."

"Ah, at last, here's the delivery: 12½ minutes late again!"

Occasionally he lets me in on his latest mathematical discovery:

"As I say, if you have three types of socks in your drawer, and the chance of taking out two of the same type is exactly half, what can you say about the total number of socks in there? I've found if you categorize the solutions you get interesting mathematical structures called quadratic forms, which are used to understand the Monster."

This is very encouraging! This biography is going to be 700 pages long! Math, the universe, genius, the Monster, socks as the source of symmetry in Quantum Field Theory! All this and Julie Christie too! The manuscript will have to be shipped to Waterstone's bookshop under armed escort and printed on cigarette paper.

Then abruptly he vanishes, mid-sentence: focus disappears, eyes glaze over. Animation, sucked back in like a reverse-action movie. Whoooo-*up*! No Simon left; just skin and absence.

A flutter of extra grin appears over his mesmerized face, and slowly fades off.

"What are you thinking?" I demand.

"Nothing."

"Just then, when you smiled like that—what were you thinking?"

"I can't remember."

"You must be able to remember! It was two seconds ago. You had a thought about something. Some story came to mind."

"No. I don't think it was anything."

"Then why did you suddenly look like that?"

"I don't know what you mean. It was just my face."

"Exactly! And your face showed you were thinking something interesting. Something I might be able to use in the book. What *was* it?"

"I don't know what . . ."

"Think! What *was* it?" I press desperately.

"Hnnh, aah, oh dear, please!" Simon gasps, dropping his duffel, thrusting up his head and gulping air like one of his mackerel. "Please! I'm not feeling very well. Let me explain in my own time. Oh dear!"

The book will be fifty pages long, enjoy a limited circulation among friends, and be printed at a photocopy shop in the Business Park.

I'll die a drunk in Deal.

To take my mind off unhappy thoughts, I return to my favorite subject. "Biography, you see," I announce, sitting cautiously on the edge of his food-stained mattress, while he drags dustbin bags of maps and timetables up and down the Excavation, "biography is about how an author chooses to present what he takes to be the facts: how he opens the story; what mood he picks for his closing line; where he places his adjectives, his quotation marks, his silences. Every biography is therefore a portrait of a relationship between author and subject, not just about the subject. You agree that mathematics is not about numbers per se but about the relationships *between* numbers?"

Simon squints at this fact, well accepted by every mathematician in the world, with a puzzled look. He never likes to agree to anything when it comes as a bald statement.

He never likes to agree to anything at all when it comes from me.

Biographies of famous people are different. Famous people are in the ring punching and jabbing every day already; you're just another little scrap on the way. It takes only a month or two for a book about a famous person to feel as if it sagged onto the shelves a decade ago.

But the unknown, I think to myself, feeling suddenly that my argument has become too precise and clever, you lock them in your image, and out of their own, for life.

You've got to be careful with a responsibility like that.

"Grief!" I shout. "You say you suffer grief? What are you talking about? Excuse me, Mr. Four-Story-House-Without-a-Mortgage, Mr. No-Need-for-a-Job, Mr. I-Spend-All-Day-Going-on-Holiday-Jaunts-and-Pretend-It-Counts-as-Campaigning-for-Buses-and-Trains: what *exactly* is it that you have to grieve about?"

It's a cliché that mathematicians are over the hill by their mid-thirties, but often it's not loss of mathematical intelligence that weakens their ability but loss of focus. They meet someone luscious in the university canteen, get married, shackle themselves with debt, find they're suddenly pushing a stroller containing twins; or they decide (surprisingly common among aging genius mathematicians) they haven't done enough for the health of the world and stumble off to fight American Imperialism. Love, financial worry, a surge of community spirit: the mathematical concentration is shot, and before these ex-geniuses can get it back, it's time to retire and die themselves.

Simon says that in his case, it was grief.

"Grief!" I marvel again. I feel almost betrayed. He has misled me. Hasn't he known for months that I've been intending to subtitle this book "The Biography of a Happy Man"? What does he mean, *grief*?

Simon stares around the room of debris and timetables, and his shoulders collapse; there are many miles to go before it is clean.

For a moment my mind is also oppressed. Ask Simon about his past, his family, his plans for the future, his reaction to the fact that his neighbor has just drowned under a canal boat and been brought back to life—and he burps, makes elephant yawns without putting his hand over his mouth, dashes out of the house on long, sweaty journeys to God's Blessing Green in Dorset. The only things about which he's at all interested in emitting something other than one of his five grunts are these timetables. And look at the state of them!

Once upon a time Simon must have been tidy and concerned. In the back room, one wall is lined by a floor-to-ceiling wooden bookcase on which everything is arranged with startling care: pamphlets, timetables, catalogues, manuals, guidebooks, route maps, almanacs, directories—where big enough, they're set upright alongside his murder mysteries and chess books; otherwise, they're lined along the bottom two shelves in color-coded boxes. Old train tickets (in green boxes) from the 1960s, his student campaign letters (yellow) from the 1970s onward, begging local politicians to improve public-transport provision.

But after 1985, shelves still only four-fifths full, the filing takes to the floor: a deluge of paper and books. And there is another difference: the catalogues on the shelves are all the same design, tend to be printed by one or two central-government bodies and are thicker than the booklets on the carpet. The floor slosh is still sour-milk-colored but glossier, issued by dozens of private companies, referencing hundreds of different regions of the country. After 1985—the time when some mathematicians claim he suffered his "catastrophic intellectual collapse"—Simon has taken to collecting thousands of timetable publications, compared to the one or two simple doorstops of before. What happened to him after 1985?

"Simon, will you please answer me?" I return to the fray. "What is this about grief causing you to lose focus on mathematics? Grief about *what* exactly?"

He sighs. I sigh. He grunts a Grunt Number Four.

"Aaaaaahh, hunnh, ugggh . . . The 1985 Deregulation of the Buses Act."

"Hello, my name's Duncan, I'm gay."

Simon has called a builder, or (since Simon has no more idea how to contact a builder than he does how to fry an onion) he's kept an eye out in case one happens to stumble out of the bushes—and one has. Duncan stands outside the front door, holding the bottom of his face in the cup of his hand as if the putty won't set.

On the street, I can see a bicycle with a homemade carriage attached to the back by an iron bar. "Duncan McRifkin's Mobile Building and Decorating . . ." The words, on a metal panel, arch across a bumpy tarpaulin in cowboy serif, black against a shadow of green. A second row of letters, made up of one word only,

stretches out thin and alone: "Service." Its curved opening S and closing e puff out at each end of the sign like bloated cheeks.

"The solution to all your troubles is on there," says Duncan, releasing his hand just long enough to point at the bike, then snatching it back.

He's come to pull the Titanic Toilet out of the floor. He's also

going to repair the bean-sized holes in my bathtub, put treads on the booby-trap stairs and install a new shower upstairs. "This is a pipe bender," he says, lugging a large tripod through the door. "Did I tell you? I'm gay."

"What's that scar?" I point at a rip on his neck that looks as if someone had chopped his head off, then stuck it back on without wiping away the excess glue.

"From when I tried to hang myself." He puts a pipe in the bender and gives it a thwack with a wrench to show me how it works.

To kill himself, Duncan had gone into the attic of a friend's house with a rope round his neck, tied the other end to a radiator, then flung himself out of the skylight window. But halfway down, it was the rope and not his neck that snapped.

"So I went back upstairs and tried again."

This time, Duncan doubled up the rope.

His luck didn't improve: again, the rope broke and he crashed into the flowerbed, his neck still whole.

Not a man to accept his incompetence hastily, Duncan went upstairs once more and hurled himself into the air for a third go. This time, however, he caught his thumb on the edge of the kitchen window, and split it open.

"Ah," he thought, picking himself up from his friend's now ruined flowerbed, "this gives me something to do." Off he went to Addenbrooke's Hospital to get his thumb patched up. It was the nurse who pointed out the splendid Chinese burn under his chin and the fact that his Pringle sweater was drenched in blood.

"I was in Fulbourn Mental Hospital for six months after that. Have I mentioned that I'm gay?"

Duncan likes to take a nap in his underpants after lunch. One day, about a week after he'd refloated the Titanic Toilet, and was about to have his siesta upstairs among the mangled shower pipes, I overheard Simon talking to him.

"Ahhhh, hhhnn, no, Duncan, I'm not gay. But I think if I was interested in sex, it would be with females."

*27 Garbage Bag Group

> I've noticed before that you like to wallow in filth.
>
> *Simon, to the author*

As soon as the word "always" is mentioned, in any context whatsoever, in rush the mathematicians shouting about symmetry. Here's a slightly different application of "always":

When Simon's cleaning up his floor, and throws out an

envelope , a duplicate 1970s train ticket

and an Indian takeaway leaflet   , it doesn't matter in
what order they end up inside the rubbish sack. It's all garbage.
It's *always* the same to him how they manage themselves in
there, just as long as they're removed from the floor, out of
sight and off to the county incinerator.

This selection of three objects in a rubbish bag must, in
some way, have "symmetry." Once Simon has put the items
inside, no matter if he shakes the bag, it's *always* three objects
inside a sack. It's the equivalent of the fact that no matter how
much you rotate Triangle or Square through their symmetry
turns, they *always* look the same. Shaking the bag changes the
order, not the contents. The order of the items is, for garbage,
the equivalent of rotation for squares and triangles.

Let's say Simon leaves the room to go on a jaunt to Spital-
in-the-Street, Lincolnshire. I sneak downstairs, pick out the
garbage bag in question, and furiously kick the thing around
the Excavation.

Beating up this bag is going to change the order of the
contents. They might originally have been in this arrangement:

But after a good thwack with my foot, they have ended up in this one:

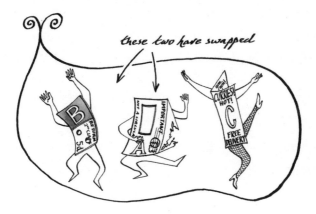

these two have swapped

The envelope and the train ticket have changed position. In other words, I've delivered a particular kick

that has interchanged whatever was in position 1 in the bag with whatever was in position 2. I've "rotated" the order:

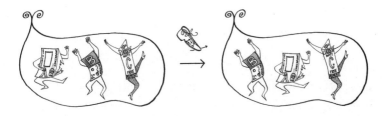

If I gave the bag exactly the same kick again, the objects in positions 1 and 2 would be swapped back, and return to their original position, i.e., the result is bugger-all. It's the garbage-

kicking equivalent of  for rotating a square or for an equal-sided triangle:

Alternatively, I could have used a different kick the second time round—say,

which interchanges whatever is in position 2 with whatever is in position 3. So:

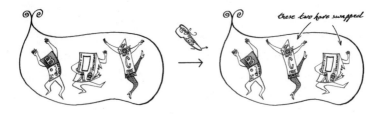

In short, followed by has, overall, produced a proper jumble, in which everything has ended up in a different place. The envelope that used to be by the neck of the bag, in position 1, has been beaten to the other end; the Indian takeaway leaflet has moved left, up to the middle; the bus ticket has gone from the middle to the neck. This is a truly sassy kick, and needs to be represented by a special shoe:

So the types of kicks we've had so far are:

| Leave bag contents alone | Swap bag contents in positions 1 and 2 | Swap bag contents in positions 2 and 3 | Make a big mess |

All told, there are six different ways three pieces of rubbish can be lined up inside a garbage bag—and therefore six different types of kick needed to get them into all the possibilities. Once you've got them all sorted out,* you can draw up a big Group Table again to keep track of how these kicks combine, just as with the regular triangle and the square:

---

*The other two are:

| Swap bag contents in positions 1 & 3 | Make a different big mess |

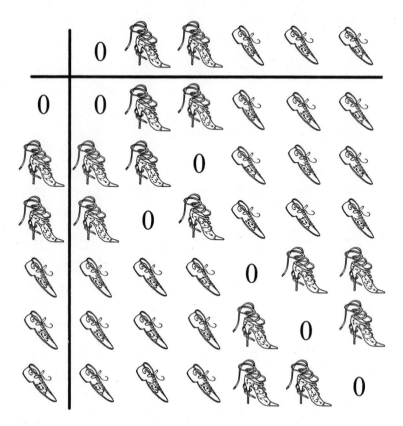

The Garbage Kicking Group

And here's the essential point—the reason for my obsession with kicking Simon's garbage. This Group Table represents exactly the same Group as the one for the rotations and flippings of Triangle in Chapter 13. All that's different is the notation.

But replace each of these symbols . . .

by these . . .

. . . respectively, and you'll find the two tables are identical. You might never have imagined that the rotations-and-flips-of-a-triangle and kicking-a-bag-of-rubbish-around-Simon's-Excavation had anything in common, but they are examples of exactly the same thing. They are just two of the day-to-day, commonplace manifestations of a particular type of symmetry. As with all symmetries, this one is embedded deep in the metaphysics of the universe. Like a mischievous Hindu god, it pops up to our world whenever the fancy takes it, in any one of a billion different disguises.

Group Theory is at last starting to show its power. It peeks under the surface of existence. It gathers our world into orderliness.

That's enough about garbage.
Next chapter, please.

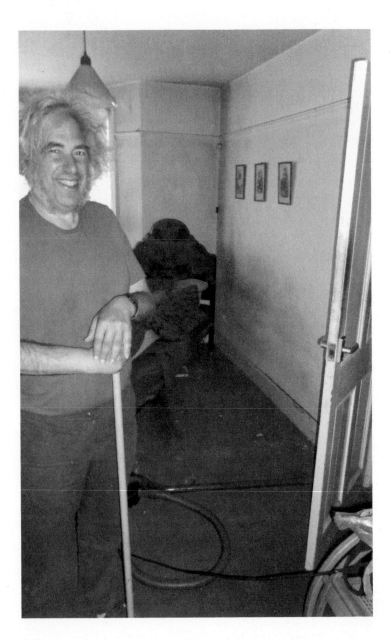

28

You must be very careful not to jump to easy answers with Simon. That is a danger. I think the subject of your last book told you that too.

Professor Conway, to the author

Simon appeared beside me suddenly and silently (as happens much too often for such a clumsy man). After a moment of fishy gasping, he dropped his bag onto the bus terminus pavement.

"Hullo."

"Hello."

"Hallo."

"Hello."

"Hnnnn."

Introductions now probably over, Simon wrenched at the buttons on his puffa jacket and tipped himself forward to squint at the seventy-two-point headline of the timetable. He was taking me to the AGM of the Sudbury–Haverhill Railway Action Committee. "Haverhill, Haverhill," he muttered, and glowed with purposefulness and anticipation.

On the other side of the noticeboard a double-decker rumbled gently. Above the driver's window, written in LEDs, close to a foot high, was "44: Haverhill."

The driver caught my eye. I nodded back. I raised my eyebrows in companionship.

"Simon, he wants to go."

"Hnnnnn, aaagh . . ." said Simon.

243

I shuddered and stamped my feet. An arctic wind sliced up Emmanuel Street and buffeted the departure screens.

"Simon, if we don't get on now, he'll leave."

"Hnnnn," said Simon, and opened the zip of his duffel. For a few seconds he appeared to be assaulting the bag, battering it from inside; then he stood up, triumphant, gripping a folded leaflet: a Cambridge–Haverhill bus timetable, scruffy, out-of-date. Simon shook it at our bus.

The driver slapped his hands back to the wheel, jabbed the ignition and rumbled off between the empty buses on either side. Into the hollow rushed the wind, picking out each crack in my overcoat.

"Hnnn," repeated Simon, gazing with satisfaction after our vanished transport. "Next one in twenty minutes."

He dropped the folded leaflet back into his bag, and bent back to the station timetable, running his fingers across a line of numbers, stroking their heads.

Simon's lived in this city for thirty years, but he still likes to confirm his exits and entrances.

On the next bus, the first thing he did was get out his map of Cambridge, open it up with a strenuous shake that biffed me on the ear, and place his forefinger at the center. As the bus moved off—left, along Parker's Piece—right, past the swimming pool—left, at the Catholic church—Simon's fingertip twisted and turned with it.

"But I still don't understand, Simon, why are we taking this service and not the previous one?"

"Excuse me, don't talk to me now, please. I am rather busy."

Past Addenbrooke's Hospital, glowing with streetlamps and bulbs to warn airplanes away from the incineration tower, up the Gog Magog hills, above the city's sparkling lights.

"This service is the Number 13," said Simon, tugging a catalogue out of his pocket. He flicked through the pages, tapped a couple of numbers on the nose, and squeezed the book back into his puffa jacket. "The other was the Number 44. This service," he repeated with greedy satisfaction, "takes longer."

On the other side of the Gogs, we plunged among the hollows, into where night had built up a forest across the road.

On the upstairs front seat of a double-decker, there's nothing to stop a man from exploding through the windshield when the driver brakes, so Simon never sits in the front seat. He places his *bag* on the front seat, himself in the row behind, then grips any rail he can reach like a fairground rider. Even on Conwy Bus Route Number 256 he wouldn't be so giddy as to try to get a good view of what's going on by sitting with his face against the glass. The 256 runs up the shoulder of Mount Snowdon, surges over the Aberglaslyn Pass, and idles down through forests and lakes to a flickering Irish Sea—Simon's favorite route "in terms of visual satisfaction."*

At Linton, the bus stopped for a few minutes beside a humpbacked bridge before entering the danger zone of village backroads. It crouched massively under the streetlight, grumbling and shaking. The town was quiet and ready for midnight, although it was barely past 7 p.m. A woman clickety-clocked out of an alley, scuttled across the cobbles, disappeared into another alley. The bus took in a breath of air, stiffening itself for the lanes ahead. Simon finished his carton of passionfruit juice, a thin trickle of yellow dribbling down his chin onto his T-shirt, and clamped his gaze and finger back on the map.

It didn't seem to bother Simon that there was nothing to see except stars as we knocked our heads between the windows

---

*His favorite route in the South is the 4 or 47 bus: Newbury–Lambourn–Swindon.

and the seat metal while the driver bashed around in the dark looking for ever more misdirected lanes to plunge down.

Simon kept his eyes fixed firmly on the paper in his lap, judging right and left turns by the sway of the bus and the sound of oaks and hawthorn hedges scraping the metalwork, inching his finger along the lines until we were back on the B1052. Then he breathed a sigh of relief, rubbed the condensation off his window, and looked out.

Haverhill is sunk in a shadow-filled valley, sodden with frozen-food supermarkets and shriveled chain stores. A long, pedestrianized high street splits the town; there are some suffocated churches; HMP Highpoint (where Ruth Wyner and John Brock, two Cambridge charity workers for the homeless who were imprisoned on trumped-up charges for, respectively, five years and four years) is up the road.

("You can add here," says Simon, "that I admire John and Ruth as prisoners of conscience.")

The "Idle Banter" page of Haverhill's local website had to be shut down because the residents persisted in using it to swear at each other. For some reason Simon wants this wretched place to be connected to Cambridge by rail link.

At the bus stop, Simon squeezed his bag between the doors before they were properly open, propelled himself past a Co-op that glowed in the night with an air of bleakness and additives, and squeezed through a narrow walkway into the pedestrian thoroughfare, once perhaps beautiful, now jackhammered into ugliness by the Planning Officer's obsession with colored concrete bricks and faux–Victorian ironwork signposts pointing to the toilets.

"This is it," said Simon, coming to a halt in front of a converted church. We peered through a pyramid-shaped window at a community noticeboard and a poster for the Samaritans. But instead of going inside, he was off again, duffel banging against his side, body tilting right to counteract the pendulum effect, bushy head bowed—more a barge than a walk—past the "Wanted for Murder" posters on the police station railings and the crowd outside the Wetherspoon's supermarket-sized pub. Where was he going? He is such an irritant to pedestrians.

"Simon?"

Into another church's yard—out again.

"Simon?"

At the edge of a little lane he paused to investigate the dark for traffic.

"Si—monnnn!"

And disappeared through a twenty-foot hole knocked into a medieval wall.

Always, after the bus station and checking the location of the "venue," Simon's priority is to identify, walk to and—with an asexual stiffening of his shoulders—penetrate the public library.

Then he feels positioned.

By the time I'd reached the book stacks myself, Simon had crammed his bag with new booklets, maps, timetables, invitations to charity rose-garden days, promotional flyers for afternoons of barn owls and falconry, recycled-paper adverts for balloon flights over Saffron Walden, pamphlets from the local-history society on the subject of Haverhill's frankly minuscule role in the Tudor wool trade; and Arriva Transport's latest handbill about burst pipes.

He lost me again among the shelves. I had to trot up and down the metal rows trying to catch glimpses of him in the rectangular gaps between *Woodcarving with a Chainsaw* and *Collect Fungi on Stamps.*

"Simon? Simon, is that you?"

For a second Simon emerged from an aisle, scooped the leaflets off a table covered in local-authority publications, then bounced back into obscurity along a row starting with *Beyond Leaf Raking*.

"*Simon! Psst!*"

His head is like a gray feather duster, at the other end of which is a little housekeeper, cross and impatient to get home.

Across the car park, two muscular spruces spread along a lawn.

I glimpsed a head bob above *Inflammatory Bowel Disease: A Personal View.*

"*Simon!*"

The fluffiness disappeared behind *Drying Flowers with Your Microwave.*

"*Simon! Was that you?*"

It reappeared three aisles to the left in audiobooks, and began to peck up and down, up and down. He was emptying an entire information-display carousel of its contents.

Abruptly Simon was beside me. His bag was covered in fresh bumps; it was bloated with new leaflets; his trouser pockets bulged with handbills and flyers about zorbing (rolling down hills inside a large bouncy ball). Wedges of brochures were squeezed behind his ears and sticking out of his socks. A small dribbling of leaflets fell from the rips in his puffa jacket and started a pile next to his shoe.

"Ahhhh, um, Alex," he said irritably, breaking off for a second from his delighted chewing of his tongue, "why are you still here?"

A mild, angular-faced gentleman, pink tie poking over the top of his V-neck sweater like a noose hastily stuffed out of sight, he patted the table and cleared his throat: "Ladies and gentlemen, good eve . . ."

"At school, we washed clothes every week," mused Simon in a noisy attempt at a whisper. "I sometimes leave it a bit longer now."

We'd arrived back at the church for the 123rd protest meeting of the Sudbury–Haverhill Railway Action Committee. The gathering space was half filled with chairs. It was not a hall. It was a widened portion of corridor that concluded somewhere better and more important, farther down. Dotted around the chairs were eight men and two surprisingly young women, both with an air of having been unexpectedly turfed out of their homes.

Reverend Hill, leading light of the Sudbury–Haverhill Railway Action Committee, held up the minutes and gave them another revolutionary shake.

"Ladies and gentlemen! Good evening! Tonight we are going to . . ."

Simon began rustling in his bulging duffel. After a moment of loud plastic rips and the sound of punched shopping bags, he re-emerged with a thick fold of paper, opened it out, heaved a sigh, and settled back in his chair.

Reverend Hill, paper still in the air, looked across and nodded, but not with irritation. Simon has been coming down on the Number 44 and making peculiar noises at these meetings for the last ten years.

"Ladies and gentlemen, the purpose of our mee—"

Again Simon dived inside the bag, rumbled around this time with the crackle of thunder, and returned to his seat with a bigger prize: a carton of Tesco UHT orange juice. He squeezed the top to open it. The liquid spurted on his nose.

Reverend Hill gave an ecumenical laugh. "Ha, ha, terrible, isn't it, modern packaging? Would you care for a tissue? No? Quite settled now? Well, as I was saying, gentlemen—and ladies—this, being our AGM, we have a considerable quantity of material to cover, so let us move quic—"

But Simon was still not quite settled. A vision of his position in the world had passed in front of his eyes—and it contained a gap. This time, the explosion of plastic resulted in a pickled cucumber. Now he was ready. His mouth widened for the opening scrunch, and the Reverend Hill began a third time.

Simon has attended a good many gatherings like this—meetings with an air of studied legality and respectful advancement, and all eyes turning upward at the stupidity of politicians. I find them intensely frustrating. If you are going to act, then *act*—smash a window, undermine reputations, slash your enemy's tires. If all you want is to whisper and reasonabilize, then do it at home with the lights off. But Simon basks in this sort of thing, and comes away with a sense of exploiting the democratic process and helping people.

The first item on Reverend Hill's agenda was about a letter "expressing concern" that the committee had sent to the local Member of Parliament.

Lo! Surprise! It had not been answered.

Careful discussion, reasoned debate, seventy-six committee resolutions signed in quintuplicate had gone into the composition of this object, since (who was to say?) some day it might become Central Exhibit A in *Haverhill's Bumper Book of Public Transport Thrills*—and the MP had ditched the pages in the bin along with his egg scraps.

The letter out of the way, the treasurer read the committee accounts: £12 in subscriptions, £22 in donations, £6.50 from the sale of postcards, minus £54 to pay for the hire of the meeting room, minus another £14 for heating. Yet somehow these pathetic sums had added up. They had £1,307 in the bank.

"That's a significant number," said Simon with a grin.

Reverend Hill turned to make some remark, then abruptly stopped and took up tapping his cheekbone instead. There were many simple retorts he could have made to Simon's maddening interruptions—"Oh, is it? And what's that, what's this latest absurdity got to do with anything, my dotty friend?"

Or, "No, Simon, not now, do for once try to keep your mind in this world, and shut up." Silence flopped into the room.

The treasurer, unwilling to be shaken from the single, brilliant object of his figures, started to bend the edges of his balance sheets awkwardly.

There is a famous anecdote about the Cambridge mathematician G. H. Hardy visiting the genius Srinivasa Ramanujan. Ramanujan, aged thirty-two, was in hospital in Putney, recovering from a suicide attempt.

"I have just come in a taxi that had a rather uninteresting number," complained Hardy. "I hope it is not a bad omen." The number was 1729.

"No," replied Ramanujan. "It is a very interesting number. It is the smallest number expressible as the sum of two cubes in two different ways."

$$1729 = 1 \times 1 \times 1 + 12 \times 12 \times 12$$
$$1729 = 9 \times 9 \times 9 + 10 \times 10 \times 10$$

Such numbers—numbers that can be expressed as sums of cubes in a variety of different ways—are now called Ramanujan-Hardy numbers, or taxicab numbers.

There is no known fundamental importance to these things—yet. For the moment, they simply possess the sort of balance and surprisingness that please the artistic sensibilities of mathematicians, and they come with that nice anecdote attached. At the same time, that doesn't mean they aren't worth profound study. They appear in books on Number Theory and are the subject of earnest articles in computing journals, because the amount of calculation required to find them requires huge quantities of computing power. You can never be certain with mathematics. It is possible that one day they might explain how whales migrate, or galaxies implode.

So, when Simon said to the baffled militants of the Sudbury–Haverhill Railway Action Committee that 1307 was a "significant number," I couldn't help recollecting this story about Ramanujan and G. H. Hardy and the taxicab with the "dull" number.

Instantly I pressed the record button of my tape recorder, specially brought along to capture just such a moment, which but for me, Simon's Boswell, would be lost to mathematical history; a moment that will lead to a new branch of study in Number Theory, the Norton-Masters numbers.

"The A1307," pronounced Simon, rustling in his plastic bags again and plucking out this time a pale-blue pamphlet of timetables. "Hnnn, it's the road from Haverhill to Cambridge."

29 Great Silence

Dr. Parker: To prove something "in the sense of Simon" is to do a proof without doing it at all. You just know the answer will come out right without having to do any calculation.

Me: Can you give me an example of this that I could understand?

Dr. Parker: Wooooooo! . . . no . . . no, nothing so simple as that.

Dr. Richard Parker, Simon's former colleague,
in conversation with the author

Simon has dismissed himself from his Cambridge history. He hurried about the city during these student years with such a platter-faced lack of awareness that (from the point of view of his biographer) it's possible he wasn't there at all.

"But your own history! How can you eliminate that?" I say. "Are you a person or a fog?"

"Ahhh, have you seen this?" says Simon, putting down his latest garbage bag. (There is now a clear view of carpet stretching from the kitchen to the main door.) It's a till receipt for a book called *The Messianic Legacy.* The purchase entry reads: "Messy Legs . . . £8.00."

Simon's life during the period in which he was a student and a world-leading young research fellow is like a film in which a minor character has been removed from the celluloid with a Stanley knife. Other figures start back from his hollowness, tiptoe away from his vacuity, whisper behind a back that isn't there. There was none of that noisy, gawky scrabble to establish a character for himself that makes other young men such memorable bores. Simon aged twenty was in extremis what Simon is being today, at fifty-eight: a gathering of emptiness. His old classmates were thrilled to know this famous genius was in their year; delighted to be counted among his admirers and conversationalists; overjoyed to point him out to their baffled parents when they spotted him scraping along the wall toward the Mathematics Faculty with a sequence of plastic bags on his arm (this was before he discovered the existence of duffels)—but as a man they barely remember him.

"He had this hermetic life," says Professor Bernard Silverman, now a Fellow of the Royal Society and former Master of St. Peter's College, Oxford. "At an angle to the rest of the world."

"Recently, I looked him up in *Who's Who,* and when I didn't find him, I assumed he'd finally got run over," recollects Raymond Keene, who became England's second-ever chess grandmaster. "Our main concern was making sure he got to chess matches without getting knocked down, because he was always looking at his feet as he was about to cross the road."

Lee Hsien Loong, now Prime Minister of Singapore, and the second-best student mathematician at Trinity after Simon, "feels he did not know Dr. Simon Norton sufficiently well, and it would not be appropriate for him to do this [i.e., comment]."

"But what sort of person was he? Did people like him? Who were his friends?"

"Of course people liked him," soothes Nick Wedd, who went up to Cambridge the same year as Simon, to study natural sciences. "Why would you not like Simon? What was there not to like?" For a giggle, he and Professor Silverman broke into Simon's rooms in Trinity once. "But we didn't find anything there either."

Grandmaster Keene rings up with a new recollection. If I bring him a ten-shilling note next time I'm in Clapham—a used note, not a fresh one—he will re-create it for me.

"As Simon walked around," explains Keene when I arrive with the money, "he used to pester the note, like this:"

Keene also contributes the astonishing information that Simon wore a suit during these years.

"Don't make me laugh!" I reply, outraged. "He wouldn't know how to get into a suit. He'd put his feet in the armholes. His head would appear out of a trouser leg."

"And a tie," insists Keene.

Simon too looks dubious about this piece of information.

His brother Michael remembers that as a child Simon invented an idea called Vortex Theory. According to Vortex Theory, one step in the wrong sartorial direction—e.g., buying a new pair of trousers when there are still two days left in the old ones before the police file indecency charges—and the Vortex will get you. A second later you'll be swirling down Saville Row in a frenzy of designer suits and Gucci tiepins.

(*Simon:* "'Savile', not 'Saville'. I may despise the products of *Savile* Row, but not so much as to want to misspell the name.")

However, Keene's evidence does explain a mystery of the Excavation: why there is a clothes cupboard in the front room (wedged against a groyne halfway along the south wall) containing, sodden with mold, three jackets and a dressing gown. The thin filaments growing along the fibers of the gown are different from the fried-egg splatters sucking on the jackets.

At the base of this cupboard, a swell of floor clutter has washed in and left a puddle: three shoes and an Asda bag of Norwegian swimming brochures featuring young ladies.

"As I say," continues Simon, inspired by this flurry of my historical investigation, "I do remember one other fact pertaining to my habits of the period . . ." He stretches his back as though limbering up for a shocking revelation: "I was always a great Twiglet eater."

Then he returns to sorting through a pylon of colorful leaflets by the cupboard. "I think this pressure you are putting on me may be creating a domino effect. One memory seems to be coming quickly after another now."

While at Cambridge as a student, Simon also developed a dislike of cows "when of the female sex."

His habit of scraping along walls as he walked was not because he was trying to keep in contact with it; he was avoiding the middle of the pavement, although he can't explain why he should have wanted to do this.

He adopted a new word: "ooze."

"What did it mean?"

"Agggh, hunnnh, I don't know."

"Ooze": it was between an interjection and a sigh.

"Where did you make it up from? The River Ouse in Yorkshire? 'Ooze' as in what mud does?"

"I can't remember."

"Well, how did you spell it?"

"I never spell it."

Thumping down medieval wooden stairs from his college rooms at every opportunity, past the students lounging about on the lawn with Cava and strawberries, Simon hauled his plastic bags through streams of cyclists on King's Parade—and dropped himself into the first bus at the Drummer Street terminus that would fling him into distant Britain: the glens of

Scotland, the tin fields of Cornwall, the glittering lakes of Cumbria.

After his first term Simon rushed back to London and calmed his jitters with a twelve-mile Ramblers Association walk around Meopham. He still has the leaflet stored in a box file, on the oddly tidy shelves in the back room of the Excavation. "On behalf of British Rail," it announces, "we apologise to the participants on the last trip who had to leap out of the train next to the Live Rail and remove a fallen tree."

Simon joined an excursion group: the Merry Makers, a British Rail club run by an energetic fellow who claimed at the top of the letter that he was the Area Manager of Watford Junction train station—but signed himself off at the bottom, in a mumble, as the Commercial Assistant. Saturday Sprees! Holiday Previews! Spring Tours! Pied Piper Seaside Specials! Simon leapt on them all.

"The two things," he says, "that I would recommend to anyone who is lonely: politics and public transport."

Suntan Tours! Autumn Tours! Mystery Tours (Mini, Midi or Maxi)! "Hello, you may already be a 'Merry Maker' and I am sure you will not be able to resist becoming a 'Happy Wanderer' too!"

It was during this period that he developed his distinctive technique for introducing himself to strangers.

He locates a bunch of people he likes the look of, looms at the fringe of their conversation until his quick mind spots a link between a passing topic and a place he's visited recently, then he whips out his wallet, extracts the relevant ticket and jumps aboard. Lo! He's with you, rattling along on the conversation.

"Haaa, hnnnn, have you seen this? *[Hands round a travel docket from, say, Pershore.]* I thought you might be interested because you were talking about Anthony Powell and earlier mentioned the village of Wyre Piddle…"

Simon looks semi-idiotic at these moments. But he's the opposite—while you're presuming that he's lost his marbles, he's waiting for you to spot the three or four links connecting Piddle and Powell. Piddle is next to Pershore railway station, Worcestershire; last week Simon made the journey from Piddle via Pershore to Frome; Frome is where Anthony Powell had his house "The Chantry" and wrote *A Dance to the Music of Time.*

It's quite possible that Simon has read a novel by Powell (despite Simon's insistence that he dislikes all serious literature), and could have joined in with a sharp comment about one of the characters, but if he can get to the topic this roundabout way, on public transport, he feels vastly happy.

When I have a party in my rooms, Simon stomps upstairs bringing a very thick train timetable—the thickest possible: one for the whole of Europe or America. Most people think this is because he can't bear to tear himself away from his closed-off world of public-transport obsessiveness. In fact, it's to give him *more* ways to escape his closed-off world: it's in case his ticket collection isn't enough. He wants to be able to use all the information in the book, as well as all that in his wallet, as possible links. As he walks around the room, lurking at the edges of the groups, listening out for conversations that interest him, he keeps one finger stuck among the pages, halfway through, so that the moment he hears a cue word he can snap the book open, spin through the sections, and have the relevant timetable up and pushing aside everyone's wineglass before the topic has moved on.

"Aaah, uugh . . . have you seen this? *[Shows a timetable to Heathrow, interrupting a conversation about glass eyes.]* As I say, that reminded me that I once gave £10,000 to the man who Super-Glued himself to Gordon Brown."

"Thank you, Simon."

"AAAhnnh, as I say, his name was Dan Glass. He is a founder of the activist organization Plane Stupid, which protests against the expansion of Heathrow."*

The winter term of Simon's second year at Cambridge arrived: the Garden House riots. A thousand students drummed on the windows of this local fancy hotel and invaded the kleftiko course during a visit by a delegation of fascist Greek generals. Nine arrested, lecturers reprimanded—the most famous student political event of the decade. Where was Simon? On a bus.

("Alex, I would like you to add at this point: 'Today Simon's main association with the hotel is disgust at how the view from the riverside walk opposite was spoilt by a car park.'")

Spring term showed up: rain, gray skies, late nights at the library, brandy and muffins in the Junior Common Room. Simon? At Leicester Central train station. It's his eighteenth birthday: February 28, 1970. He's standing outside in the

---

*Campaigner "glues himself to PM"

A campaigner against Heathrow Airport's third runway has attempted to glue himself to Gordon Brown at a Downing Street reception.

Dan Glass, a member of Plane Stupid, was about to receive an award from the prime minister when he stuck out his superglued hand and touched his sleeve.

Plane Stupid says Mr Glass, from north London, then "glued his hand" to Mr Brown's jacket as he shook his hand. But Downing Street said there had been "no stickiness of any significance."

Plane Stupid recently gained publicity by mounting a protest on the roof of Parliament. Spokesman Graham Thompson said Mr Glass—a 24-year-old post-graduate student at Strathclyde University—had smuggled a small amount of glue through Downing Street security checks in his underwear at about 1700 BST. . . .

Mr Thompson said his organisation was attempting to make Mr Brown "stick to his environmental promises" (news.bbc.co.uk/2/hi/7520401.stm).

The award Dan Glass received is funded by Simon. Every year, very quietly—it's appropriate to reveal it in a footnote—Simon gives £10,000 to the Sheila McKechnie Foundation to sponsor their Campaign Award for transport activism (www.smk.org.uk/transport-2008/).

desolate cold, wondering what to do because the connection he'd planned to take to Rugby Central, as a birthday present to himself, has been shut down.

Simon can remember buses and trains he took during his time at Cambridge with astounding precision; it's everything else that's vanished: "I started on the 6:33 train to St. Ives, a line that was soon to close. At St. Ives I got on a bus to Huntingdon, then by train to Leicester via Peterborough. I spent most of the day looking round the city but can't remember anything about what I saw." Summer came: silken, braless string tops and miniskirts—oh! Summer, in Cambridge, for boys! . . . Simon? Barrow-on-Furness. Eyes on toes. Stuck in a train tunnel.

He drank orange squash. The chest under the mantelpiece still has a drawer packed with empty Tango bottles, twenty years too old to return for the 2p deposit.

He went on the Masochists' Special. Simon's still got the clipping, pressed tidily inside a B.D. (Before Deregulation) box file in the back room.

Why did he find these trips necessary? What did they have that was better than college life? What could he do on them, aside from pay ten pence to sit above a bus diesel engine with farmers' wives and shrieking schoolchildren? Was he being metaphorical? As in, buses depend on timetables; timetables depend on numbers and patterns: therefore, taking buses equals traveling through math? Was he being *that* weird?

"Oh dear!" Simon complains. "Is joy such a hard concept to understand?" Hadn't I [he points at me] admitted that I used to pick up paper squares that somebody had spat on, and called it "stamp collecting"? Other people like to drop out of airplanes at 10,000 feet, dangling from colored bedsheets. Simon zipped through the countryside in metal worms.

But . . .

261

400 rail fans take masochists' special

By Wendy Hughes

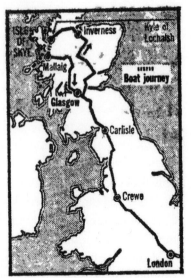

The 1,250-mile excursion

FOUR HUNDRED travellers set out last Friday evening from London on a marathon 39-hour round trip railway excursion. The 1,250-mile " Grand Highland Circle " Merry Maker trip via Inverness and the Kyle of Lochalsh, ends back at Euston at 10 o'clock this morning and is the longest excursion trip so far organised by British Rail.

One of the passengers said: " I suppose you might call it a masochist trip. We know we won't sleep for two nights but it's worth it." For the 400 passengers aboard only 86 sleeping berths were available.

Tickets for this weekend's trip, which includes a three-hour steamer trip from the Kyle of Lochlash to Mallaig, cost £7.80 each, with an extra £4.20 for the lucky few who got sleeping berths.

The ' clubroom " is the buffet car which is open all the time and where the regulars have a little sing-song when the others have got their heads down, and where they gather to reminisce: "Were you on the trip to Wales when we went on the steam train and the boiler blew up? That *were* funny, weren't it? "

" Yes, and the time the train forgot to pick up passengers at Welwyn Garden City and we had to back for miles on the main line ! "

" And do you remember when the train ran into a herd of sheep? The old guard got out and counted up the bodies."

" I've given up the mystery trips. They all end up at Margate."

The most regular and colourful customer on the trip is 72-year-old Ted, a retired tramp, dressed in a bright-red waistcoat, red tee-shirt, and blue jockey cap. " I'm off to Barmouth next weekend. I'm booked on eight trips till the end of June. It's lovely, darling," he said. " I'm going to places I would never have gone travelling. I used to do the South Coast."

" If you can take a British Rail pork pie for breakfast—you can take any-thing."

Clipping abridged by the biographer. Source unknown.

Please! Go away, biographer! That's enough! How much more do you need to be told? Look at Loch Lomond and Glen Coe swish past! See the valleys of Devon plunge and weave! Watch out, stand back! Here comes Simon, aged twenty-one years, seven months and four weeks, out of Llandudno on a Number 47, racing up Snowdonia, through Llanberis Pass.

Wheeeee! Over the hill! Rushing to the embrace of Colwyn Bay; looking out to the Celtic and light-splitting sea.

"Nevertheless, Simon," I keep at him, "if day trips were the only way you enjoyed spending your spare time during this period, why don't you remember anything about the scenery or stories that took place? Why nothing except the bus numbers and connection times?"

"Oh dear, oh dear, does everything have to have such a ridiculous reason? I liked taking trips, that's all. Why ruin it with words?"

One of two box files of tickets from the early Cambridge years, found in the back room.

30

I do not wish to commit myself to meeting you because I now
have less than 2 weeks to use up 4 return train tickets . . .

Simon, email to the author

Trinity College, 1971—Simon's third year. Simon has a vivid
recollection. He'd been moved into scholar's rooms off Great
Court, the heart of the college: mullioned windows, oak
floorboards once trod by Newton, G. H. Hardy and
Ramanujan; a fire crackling low in the grate . . .

One night, "a machine was on in the street" outside his
window. It was a pump of some sort, clearing up a water leak.
Simon thought it would cease at midnight.

But it didn't.

No other occurrence during the eighteen months that
followed disturbs the sleep of his memory.

His next room was a tenancy at 71 Jesus Lane. Here he
blinks awake a second time: he remembers an argument he
had with "some foreign temporary tenant."

"A fight? A proper fight? Goodness gracious! What a thing!
A real smash-up? Fists and punches?"

"No. A discussion about science fiction, but it's unlikely you
could discover further details, because I've forgotten the man's
name."

"What about your mother? Your father? You wrote to
them?"

"Hnnn, aaaah, oh dear. Uuugh. Why should I?" Simon protests, mumbling and catching his head. "I never had anything to say then either."

He can't be entirely unsentimental about this time, however. In a prominent position, above the piano, are two colorful prints showing Simon's Cambridge colleges. *Quadrangle of Trinity College* is a view of Trinity chapel, in the main college square: vast and stolidly ornate. In the middle is a water well. An absurd coronet of Gothic masonry has buckled the supporting stonework like jelly.

Quadrangle of Trinity College.

Jesus College, where Simon got his first fellowship after finishing his PhD at Trinity, is pictured from Jesus Green, the park immediately across the river from where Simon and I live now. Windows tinkle about the top floors of the building. The gate block is stooped and bronchitic. There is a feel of approaching rain.

"Where did you get those?" I asked one morning, when I was in the back room sifting through his box files for amusing tickets or leaflets. I was surprised. I wouldn't have thought he'd own such soppy pictures.

He pushed among the floor slosh, and stood in front of the piano, squinting at the prints.

"Ahhh, hnnn. I don't know," he concluded after a long consideration. "I've never seen them before."

View of Jesus College, from the direction of our house.

"Never was a man whose outer physique so belied his powers," said Francis Galton about the nineteenth-century prodigy and Group Theorist Arthur Cayley. "There was something eerie and uncanny in his ways, that inclined strangers to pronounce him neither to be wholly sane nor gifted with much intelligence."

Arthur Cayley

Simon Norton

Arthur Cayley (1821–95): lawyer, mountain-climber and watercolorist. He published more than 900 mathematical papers and books.

"Some people might say the same about you, Simon," I observed. We had left the house and were crossing Jesus Green, Simon limping speedily, me straining alongside to keep up. I'd been reading a biography: *Arthur Cayley, Mathematician Laureate of the Victorian Age* (600 pages, 2 lbs. Most exhausting).

"In your upbringing," I continued, "you and Cayley were just like each other."

"I have a particular partiality for 'long division' sums."

1. No friends his own size.

...sics, Chemistry, Theology, Mathematics, Greek, Latin, History, Frenc...

2. Insatiable memory; relentless desire to read every book he could lay his hands on.

"Except literature," protested Simon as we walked down Trinity Street. "There was a time at my prep school when the headmaster forced me to stop reading maths and chess books, and said that I had to read ten novels instead. But it didn't do any good. Incidentally, it was in that lane that the-pump-that-did-not-stop was."

3. Up to Trinity College, Cambridge, two years early, on a full scholarship.

"Although, unlike Cayley, you'd also taken a first-class degree at the University of London by that age," I added, "won three gold medals in International Mathematics Olympiads, and had your hand shaken by the Queen, the Duke of Edinburgh and, for some reason I've never quite understood, Shirley Williams."

"Yes, and there are a lot more mathematicians now than there were in Cayley's day, so there is more competition," agreed Simon, but not in an immodest way.

4. Achieved stellar triumph at the final examination.

After three years of study and five days of continuous testing, during which crowds of cheering admirers gathered outside the examination hall, Cayley was top student of the year and had his portrait engraved and distributed throughout

Cambridge. He'd finished the final, three-hour paper in forty-five minutes, then hurried straight to the college library and taken out Aristotle's *Politics,* Lazare Carnot's *Géometrie de Position* and Madame de Sévigné's gently racy French letters.

Cayley's nearest rival suffered a breakdown.

Yet Simon did better. According to rumor, he scored fifty alphas in his finals exam.

For the Cambridge grading system, two alphas gets you a third-class degree; five, a second-class; twelve, a first. Fifty not only made Simon first in the university but was the highest score in the history of the Mathematics Tripos—*ever.*

To this triumph of university top-ness, with which other men, with a quarter the success, revolutionize history, dominate governments, win Nobel Prizes and become infinite plutocrats, Simon has added an exquisite icing of his own: he has forgotten all about it.

Simon views his brain as a piece of clothing.

Everyone needs a pair of trousers—they can then walk to the bus station without getting arrested. Whether the trousers are Prada or Primark is unimportant. Same with brains: Simon's brain being better or worse than the next person's isn't the interesting point. What matters is, does it serve its function, which is to enable Simon to gallop about at the frontiers of mathematics? It is a fact: it does. Nothing else matters. As far as Simon is concerned, his triumph in the university exams might as well never have occurred.

"But you at least know that you got a first-class degree?" I say, beginning small.

"I would expect *that,* yes." But it never occurred to him on announcement day to run to Senate House and join the slavering undergraduates stretching over each other's shoulders to find where their names stood in the list of passes.

To any genius, degree-level mathematics is like spending three years in intense private study of a foreign language, then being tested with a phrasebook—knowing how to say "Hello," "Goodbye," "Can you please put a spare tire on my rental car?," "Take your hand off there or I'll scream," and "Where's the toilet?" Degree exams in mathematics aren't harder than that.

"So you don't know that you came top in the whole university?" I pressed.

"It is possible that I did," he confirmed with indifference. "But I do not remember it, no." We'd just hit a dumpling of German tourists. Abruptly remembering something he wanted, Simon dropped his bag on the pavement. The Germans guffawed and shuffled their beefy legs to clear a space as he yanked at the zip.

"As I say, I am not saying it is not true," insisted Simon, searching inside the bag for a packet of Bombay mix he thought he'd spotted skulking there last week. "It is possible. I am saying that I don't know it."

Simon is not modest or immodest, just as he is not vain or unvain. That he was, at that time, one of the most promising mathematicians in the world is just a fact, and what's modesty or immodesty got to do with a state of undeniable correctness? A fact is a fact. It's got no need for affect. Simon does not know that he graduated top in the university, or possibly in the university's entire history, because a) it has no relevance—unlike mathematical formulas or bus connections—to the smooth functioning of the universe, and b) after securing him a place for the next year of study, it was of no further practical value. Like Sherlock Holmes, who refused to remember whether the sun rotates around the earth or the other way round, because it was unnecessary clutter in his mind, the fact that Simon was once a prodigious passer of mathematical exams is of no interest to him. What mattered was whether or not he could continue to research mathematics after he had taken the exam. Once that fact was established—yes, he could—everything else about such teenage cockfights was unimportant.

Most mathematicians I've interviewed have managed to hint at how clever they are, and where they rank in relation to other well-known names, within ten or fifteen minutes. Better than Curtis; not as good as Thompson (who could be?); on a par with Conway. It's a tedious but apparently necessary phase of the conversation, and a good opportunity for me to glance at my text messages. Simon never does this boasting. To him, it would be like standing at the top of Everest, gasping, awestruck, at the frenzied splendor of the spectacle—and pointing out that your climbing boots had cost twenty quid more than anybody else's.

I worry sometimes when I talk to Simon bluntly about his failures; but it doesn't seem to have any lasting effect.

"So where did it vanish, Simon? Why did Cayley go on to become an immortal in the mathematics of your subject, Group Theory, whereas you, who had even better intellectual advantages—you have become a man who scuttles about on buses to Woking looking at statues of Martians?"

Munching Bombay mix delightedly, his right hand plunging and emerging from the bag, Simon crossed into Silver Street.

The pavement is narrow and was crowded; buses and lorries crushed past the curb, their loads tilted by the camber of the road, rasping along the first-floor walls above us. We huddled our backs to the wall, waiting for a gap, then shot across the traffic, through a carriage arch, into a drab and silent car park. The tarmac had settled like stiff carpet on rubble underneath. Unconsciously, I put on a teacher-like stoop and gripped my hands behind my back. This is where the Mathematics Faculties used to be housed, before they were promoted to a multimillion-pound glass-and-brick building shaped like a half-buried spider, with indoor computer-controlled microclimates, on the outskirts of the city.

Simon's department—called the Department of "Pure" Mathematics, which means math-for-math's-sake, and not for the convenience of physicists or engineers (which is called Applied Mathematics)—was squeezed into a brick warehouse beside this parking lot.

Simon marched ahead, away from this puddle of memory, his mood obscure. We passed the weir onto Lammas Field. Across the river was the Moat House Hotel gym. Behind the large glass panels, grapefruit-bellied men were pedaling wheelless bicycles; grunting up weights that crashed instantly back on their heads; thumping along electrified paths that kept them stationary. University academics go here not to get fit but because it's so expensive and unpleasant that it's one of the few places you can get away from students.

A remarkable coincidence occurred outside the University Graduate Center: a man rushing past—white hair, close-cropped, smooth face—stopped suddenly, bent forward, head slanted in query, and ventured a hand into mid-air.

"Simon? Simon . . . Norton? It's been almost fifty years!"

Simon swiveled, looked at the man, stared at his bag, looked at me . . . and chuckled with embarrassment.

"Alan . . . Dibdin," urged the man.

"I know," snapped Simon.

"We were at prep school together." Dibdin turned to me. "We were awful to him."

Simon smiled amiably, and chuckled again. "Heh, heh."

Half a century goes by and, for the second time in three months, one of Simon's prepubescent bullies from 1962 recognizes him under his cloud of gray muttonchop hair and rushes forward full of adult apology and admiration. I could hardly speak for surprise.

Coincidence makes Simon blush. He chuckled some more and appeared to count some ducks that were sitting in a puddle farther up the road.

"You're writing a biography about him?" exclaimed Dibdin to me. "I'll tell you everything. Only good things, it will be. The maths, bridge—he was good at everything. What a brain that man has! Music: always, ten-note chords." He spread his hands as wide as they would stretch and bounced them gently up and down in mid-air: "Tink, tink, tink."

"But how did you recognize Simon if the last time you met was when he was ten?"

Dibdin looked at me, perplexed that I should need to ask the question. "You can't forget Simon."

He ran through names of old Ashdown boys. "X, remember him? He's an MEP. Y, the Director of Debrett's. Z? A historical novelist. None of us pushers. We're all weak people who need structure. But this is amazing! Simon Norton! Fifty years! Me? I'm an inventor. I stay at home all day. I'm never out, pushing things. Will you have lunch with me, Simon? Please say yes."

I am Simon's only regular friend, but he is surrounded by friendliness.

In Sainsbury's, half an hour later, after we finished our walk through Lammas Field and returned across the Backs to town, Simon snatched up two packets of chicken wings, pressed them against his nose to read the labels and dropped one into his basket. "'Hot 'n' spicy.' I eat them cold."

"About those prints of Trinity College and Jesus, above your piano," I say. "How can you not have seen them before? You hung them."

He scuttled across to the crisp racks and snatched up three party packs of onion rings.

"No, I don't think I did. Aaaah, no. I don't think that is a safe assumption."

He flung all the packets into his basket, and darted through a crowd of students to crouch in front of "Condiments." This was not a supper food run. These were bus-jaunt supplies, to stash in his duffel.

"Who did hang them, then?"

"Hnnnh, aah . . . perhaps it was my brother. He and his wife I remember came up one afternoon and decorated for me—they must have hung them then."

"When was that?"

Simon looked up, adopted a calculating expression, then dropped a jar of pickled beetroot into his basket, among the onion rings.

"Twenty-six years ago."

31 The Monster

Groups can have Subgroups—smaller symmetries hiding in the larger symmetry. By looking carefully at the Group Table for the rotations of Square, it's possible to spot a Subgroup skulking inside:

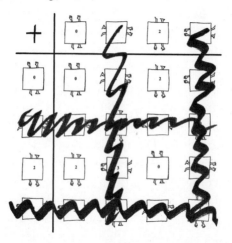

The Group Table for Triangle does not contain any Subgroups. It is an atom of symmetry. In the language of Group Theory, it is called a "Simple Group."

There are quidzillions of these Simple Groups, all sharing this fundamental quality: they have no normal Subgroups.* The last of these Simple Groups, the final piece in the periodic table of all the universe's finite symmetries, wasn't discovered until 1973 (not by Simon).

---

*The word "normal" is essential, but too big a subject to discuss here.

It is the Monster.

But what is the Monster? What set of symmetries does this gargantuan object represent? Is it just spinnings and flippings of another flat shape like Triangle or Square, only, instead of three or four edges, it has

$$808,017,424,794,512,875,886,459,904,961,710,$$
$$757,005,754,368,000,000,000?$$

This number, 808,017,424,794,512,875,886,459,904,961,710, 757,005,754,368,000,000,000, is the number of columns and rows in the Monster Group's Table. But no one knows what strange object might have these symmetries. It might be that many different objects have them. Three pieces of garbage kicked about in a bag turned out to have the same symmetries as the rotated and flipped Triangle. Perhaps it is the same in the case of the Monster Group: there are lots of different ways in which this Group of symmetries can appear in the universe, we just haven't found any of them yet. How Simon can understand the symmetries of an object before he's discovered what object might possess these symmetries I have no idea. To understand that, you have to study mathematics until your brain turns green.

One thing he does know about any object with the symmetries of the Monster Group is that it can't exist in fewer than 196,883 dimensions.

I don't know how he discovered that either. I don't even know what it means. How can something exist in four dimensions, let alone 196,883?

Simon has tried to explain the answer to me, and a little has sunk in. In a certain sense, people live and work in more than three dimensions all the time.

We taste food in five dimensions: sweet, sour, salty, bitter and umami. These are not spatial dimensions, of course, but they are still dimensions of a sort. To appreciate the full savor of Mackerel Norton, Simon's tastebuds therefore have to evaluate this mush in five distinct ways that form the basis of each tasting, from the "salt" of the headless fish, to the "umami" (one of Simon's favorites) of the monosodium glutamate in boil-the-Chinaman-in-the-bag packet rice, to the "sweet" of the flecks of freeze-dried red peppers. A complete flavor description requires five variables, which is the same as saying it exists in five taste dimensions.

We are sensual in ten dimensions. There are the five senses everybody knows about (taste, smell, sight, touch, hearing); but if you strip a human apart and study its cell structure, it turns out we have at least five other senses as well: a sense dedicated to pain, another specifically attuned to balance, a third for joint motion and acceleration, a fourth that focuses on temperature differences, and the fifth newly discovered type of sensory cell is devoted to detecting time.

A complete sensual description of Simon in the act of eating Mackerel Norton therefore needs at least fourteen dimensions (five of taste, plus the nine other senses), because each sense will have a value during the masticating process, and only by taking them all into account can the biographer reach full understanding of Simon's mesmeric joy.

This doesn't explain what it means to say an object exists in 196,883 spatial dimensions, but it starts to give the mystery a little structure. The idea that, at least in non-spatial terms, I calmly deal every day with situations taking place in four, five or ten dimensions lessens the quaking in my shoes.

For the last thirty years Simon has (in 196,883 dimensions) chased the object that has Monster Group symmetries in and out of learned journals, across campuses in England, the

United States, Canada, Germany, up whiteboards, down blackboards, jumping at it over seats at international conferences, shouting at it on the phone, pestering it around his computer keyboard and down Internet wires, into cyberspace. Even in bed Simon doesn't relax: he bounds after the terrifying thing through symbol-speckled dreams.

When Simon does not come home at night, it is because he has not gone to sleep at all. He has felt the tremor of the Monster's shadow outside his office door and jumped up to sprint after it down the Mathematics Faculty's miles of neon-lit corridors.

Or it is because he is on a bus to Pratts Bottom, not thinking about it at all.

Each time Simon snatches a piece of this prey—a clip of whisker, a splice of scale, a stray slash of mathematical eyebrow—he marks up the result in pinched handwriting that looks as if ants have crept under the surface of the page. He believes that exposing what object/objects have the Monster Group's Table of Symmetries will be among the most important discoveries of modern times.

Here's another way to illustrate Simon's puzzle with the Monster:

Simon has never heard of the man who discovered the Four Rules of Groups. These are four very simple descriptions of what every Group Table must possess in order to be a Group Table. Even a person who has failed every level of school mathematics can learn them. They are the mathematical definition of a Group. They are the equivalent of observing that a book must have pages, a spine, and two boards (material irrelevant) to make a cover, or else it is not a book. It is a scroll or a tablet or a manuscript. You'd think that a portrait of the man who first wrote these rules down would be tattooed on the forehead of every Group Theorist in the world.

"He's not important," says Simon. "I will do my best to forget his name as soon as possible."

"Not important! He codified your subject! Imagine a physicist who looked as empty-headed every time you mentioned Newton. Or a philosopher: 'Aristotle? Who he? Oh, that geezer who married Kennedy's wife?'"

"Hunnnhh, you mean widow," corrects Simon.

To me, for Simon not to know the name of the person who chipped the rules of his subject into stone is unthinkable. And it gets worse.

"I don't think these rules help you to understand Groups. They don't mean anything to me. Can we change the subject now, please? I don't like to think about it."

What Simon means when he says he doesn't "understand" the four trivial rules of Group Theory is that they give him no sense for the *potential* and *feel* of the subject. When he gets excited about these themes, essential to every good mathematician, I see Simon as the one-year-old again, playing with his pink and blue bricks, plucking patterns out of chaos.

To understand why Groups exist, what their beauty is, and how they can be turned into Monsters, forget rules, cries Simon, concentrate on your *senses*. Think of the subject *aesthetically,* develop *empathy* for it, use your *intuition.* All the touchy-feely language we would employ to characterize a good artist, Simon uses to describe good mathematical ability.

Discussing the four simple rules of Group Theory is, to Simon's mind, a waste of time, because they have no interpretative element to them. All they do is tell you something tiresome and mechanical: any Group Table that satisfies these four simple rules is a Group.

If so, bingo! The elements of such a table will form a Group and therefore describe a type of symmetry. But what type of symmetry? What kind of object has this symmetrical set of operations? The rules don't tell you that.

And this is Simon's difficulty with the Monster Group.

He has somehow acquired the Group Table, and can prove it is a Group because it satisfies the four simple rules. But he can't figure out what object or objects might have this set of symmetries. It is as though he's watching a complex swirl of light made by something dancing madly with a fluorescent tube in a blackened room. He can document all the twists and leaps the fluorescent tube makes, and therefore work out what the object holding it can and cannot do. But ask him to describe *what* this invisible contortionist is, and he looks as blank as a lake of cream.

("I'm not sure I want this comparison to cream. Cream is one of the foods I strongly dislike. I also dislike slicing bread.")

How can that be? He's got

$$808,017,424,794,512,875,886,459,904,961,710,$$
$$757,005,754,368,000,000,000$$

facts about the Monster's basic symmetrical properties, and

$$808,017,424,794,512,875,886,459,904,961,710,$$
$$757,005,754,368,000,000,000$$
multiplied by
$$808,017,424,794,512,875,886,459,904,961,710,$$
$$757,005,754,368,000,000,000$$
$$= \text{God-only-knows-what-number}$$

details about how these properties interact with each other (i.e., its Group Table).

How many more pieces of information does the greedy man need?

Perhaps just one.

32 *Atlas*

What do you mean, my genius vanished? That's the first I
heard of it.

Simon

The office where they worked was called Atlantis.

Magnolia paint frothed damply on the brick wall; the
middle of the room was punctured by a pillar; the window
frames were made of metal and shook in the wind. Atlantis
was housed on the second floor (or was it the third? Simon's
uncertain) in a converted book warehouse. From the southwest corner of this building you could just see down to the
river and punts, and, in a slice of park, girls stretching
themselves along the grass, looking firmly away from
mathematics.

Conway, Curtis, Parker, Wilson and Simon: Atlantis
threatened at any moment to sink under the weight of paper
these five mathematicians generated. Articles, books, abstracts,
treatises, napkins, paper tablecloths, backs of envelopes, torn-
off corners of cardboard boxes, reams of sprocket-holed
computer printouts—they soused the floor, suffocated the
three office tables, bubbled up the window, drenched the door.
Every day, more paper poured in—postcards from India,
summaries from Novosibirsk, cuttings and marginalia from
Honolulu, MIT, UCL, NYU, Beijing, St. Petersburg, a carbon-
copy manuscript from the University of Birmingham (always
good for mathematics), *pensées* from São Paulo, scribbled
notes of phone calls to Rome.

Conway, Curtis, Parker, Wilson and Simon were producing an almanac of Groups without normal Subgroups. Conway had had the idea in 1970: to gather together all known information about the different atoms of symmetry—a book of foundational wisdom. An atlas of symmetry.

It would take him, he thought, until 1973.

Conway would have to confirm every piece of known information, fill in thousands of gaps in the record, and dismiss all idiocies that had crept in when he hadn't been in charge.

1974, at the outside.

In the end the project covered a fraction of the original idea, involved hundreds of mathematicians sending in contributions from around the world, and took fifteen years.

Conway, Curtis, Parker, Wilson and Simon were so different in character that nothing except mathematics could have kept them sitting together in that dreary, oppressive room.

Richard Borcherds, winner of the most prestigious award in mathematics, the Fields Medal, remembers standing in the faculty tearoom one day, discussing with another mathematician an ornate calculation he'd been working on for several days. At one point, Simon knocked past and overheard the conversation. Before he was out of earshot, he'd solved the problem and called back the answer over his shoulder.

Among a select group of mathematicians, Simon Phillips Norton's status as a solver of long calculations of filigree delicacy is mythological. This is not the same as the conjuring tricks you read about in the papers, in which a schoolboy splurts out the answer to 987,654,678 x 770,645,321 in two seconds, or recites π to 30,000 places before you can tie your shoelaces. Those are unimportant, mechanical skills. Simon's genius also isn't the wild, Picasso-like brilliance of the world's greatest mathematicians, who bolt together ideas that no one had previously imagined could be united. His ability is a precise, rigidly circumscribed, top-hat-and-cravat sort of genius. It's a Nicholas Hilliard, exquisite miniaturist talent, lying a long way from the paint-by-numbers clunking of the π chanters, but still at an angle to the chaos of frontier mathematics. It's as though Simon shares his mind with a god who has a passion for making lace.

Solutions frequently appear to Simon without thought or questions about correctness. They appear—insofar as he can explain it at all—in the same way that hunger or lust or revulsion might appear to the rest of us.

Curtis, Wilson and Parker are all good mathematicians, but not in the same league.

Conway was Simon's equal—and almost his opposite. They have contradictory temperaments. Conway is ebullient, teasing, garrulous, full of joyful and moving anecdotes, effortlessly eloquent, with a fine poetic sensibility (it was he who came up with the name Monster), and as arrogant, and eager with women, as a peacock.

Larissa Queen, a mathematician from Volgograd who became Conway's second wife, remembers Simon for his modesty, good looks . . .

"Good looks!" I exclaim.

"Very good-looking. Several of the wives commented on it."

. . . and total lack of sexuality. He always had "this enigmatic expression which I described as 'I know, but can you guess?'"

I just smiled back, and that was it! I quickly realised that just because he smiles it doesn't mean that we are going to have a conversation. There was a certain gentleness about him which I could describe only as intellectual kindness. He didn't make you feel stupid, and he didn't ever make you aware that you are wrong. He would just, kind of, you know, not do anything.

And another thing I remember! Whenever somebody entered the room he'd say, "Here comes my mother!" It didn't matter whether it was a man or woman, and it was the Maths Department, so it was usually a man. You would hear several times a day: "Here comes my mother."

"Simon, OK already!" people would say. "I've had enough, could you please shut up!" Oh, he said it loud enough for a group of ten people around him to hear . . . You could be as odd as you wanted to be in that Maths Department, and some people sometimes matched Simon's oddness, but he was the most consistently, reliably odd. He maintained his permission to be odd.

And some people were not odd, and that was OK too.

Larissa often used to play Simon at the Child's Game, a form of backgammon Simon had introduced. "More difficult, but with simpler rules,"

and very bright people, very good strategists could never beat him. He told me that he used to play it with his mother. I really got hooked, and practised and practised. I was determined to beat him, because for years he had this reputation in the Maths Faculty that nobody could beat him. At last I did beat him. And, you know, he reacted with joy. After that, he often suggested we play. And every time I beat him and he lost, he reacted with joy.

"*I tried Nola's hymen!*"

Another favorite faculty game played in the Common Room was making up anagrams, especially for Cambridge mathematicians. Nola's defloration was an anagram of "Miles Anthony Reid," now a Professor at Warwick.

"*Oil thy rim, Neasden!*" Parker shouted.

"*Slime ride,* if you miss off the 'Anthony,'" exclaimed Conway.

Simon was by far the best in the department at this sort of thing. His answers were immediate, relevant, and you didn't have to glance around to see if there were any children in the room before he said them. Miles Anthony Reid was a geometer; he had recently been to Russia.

"*Earthly dimension!*" brayed Simon, shaking his glass of lemon squash, because he refused to drink tea or coffee. "*Lenin made history,*" he added a second later.

Another of Simon's anagram triumphs, which Conway still remembers with amazement, was for "phoneboxes," which Simon solved "as if with no thought" before you can reach the answer given in this footnote.*

Simon's attitude to mathematical problems was the same as it was to board games—what delighted him was the clever defeat of a puzzle. He was never interested in who got the credit for it:

> He had the same smile when he came up with some brilliant solution. People would be working on a problem and he would just say a number, some long number. And people would continue talking, and maybe two or three hours later they would realize that this number explained the phenomenon they'd been puzzling over. Sometimes it would be hours later. They suddenly would realize what Simon meant—this number which is very brilliant and requires several very deep steps to arrive at—a whole sequence of extremely complicated thinking that other people who are the best mathematicians in the world and paid a lot of struggle and several hours to arrive at—and Simon just said it. And that would be it! He would never say "I told you so," or "Aha!" He would have this joy of recognition; he would be silently pleased. And if without

---

*Xenophobes.

reason they didn't remember what he'd said, he might repeat it. And add an additional clue. And the clue could be something else really deep—or it could be just, "Here comes my mother!"

Conway hates to lose. When I visited him at his house in Princeton he pointed out a photograph of himself on the bookcase, freshly fled from Cambridge and Group Theory, in his new office in America. It was taken a quarter of a century ago. Balanced at the top of a computer screen in the picture is a number, 15.92.

It's his fastest time—in seconds—for calculating the days of the week for ten random dates. "I'm the best in the world at it," he pronounced. "There was a period when I was briefly ill, when I became second-best in the world and somebody overtook me. But that was only for about six months. Then I got better and was the fastest again! What's your date of birth, Alex? Do you mind if I call you Alex? Ah, Tuesday."

It was difficult to keep him on the subject of Simon.

"Simon? I dunno, the life of an aging prodigy is not easy. Inevitably, you compare yourself with what you were. I mean, we all suffer from it. We slowly substitute, you know, our knowledge and memory for our native wit. I mean, I was very bright, I wasn't as *natively* bright as Simon . . . or . . . maybe I was! Er, you see, OK, so let me go off about myself again . . ."

Conway is crippled by alimony payments.

Mathematically also, Simon and Conway are poles apart. Simon is meticulous, never makes mistakes; he focuses on one particular problem to the point of burying himself in a rut over it; he's a supreme manipulator of equations. Conway makes frequent schoolboyish blunders, such as putting in a minus where he should have had a plus— a small oversight, known as a "sign error," but it instantly flings an answer from one side of the universe to the other, out of quiet correctness into blazing absurdity. This does

not make him a bad mathematician; he spots the outrage soon enough.

Richard Parker (the "Parker" of Conway, Curtis, Parker, Wilson and Simon) remembers one day in the Common Room watching Conway come back from the refreshment counter carrying a full mug of coffee. Slowly, Conway turned the mug upside down and spilled the contents onto the floor. Then he turned the mug the right way up, completed the walk to the table and sat down.

"Conway!" Richard cried. "What are you doing? What just happened?"

"I made a sign error," admitted Conway. "Instead of keeping the mug up, I accidentally put in a minus sign and turned it upside down."

Parker is a third sort of theorist. There is no one type of mathematician. They are as varied as lettuces. Conway, colorful, broad-ranging, flighty but also profound—radicchio; Simon, narrow, unforgettable, perfect in a thin field—endive; Parker, unfocused, watery, not at the same level as the first two but now and again precisely right—iceberg.

I miss Parker terribly [says Conway]. He's an ideas man. He would come in in the morning and say, "I think I can cure what's wrong with last night's maths argument—perhaps it's like this . . ." And I'd say, "Richard, that's really stupid," and prove that it was just nonsense. Then he'd go away and say, "OK, how about this?" So I'd say, "Give me ten minutes and I'll come up with a counterexample." But he was never daunted. And every 100th time . . . Once, he'd taken three cases in which something corresponded to something else and said that maybe there's a one-to-one correspondence between these things, and I said, "Richard, you're so stupid." To base a conjecture on three cases was ridiculous. So we took a fourth case, and a fifth . . . I was getting a plane to New Jersey the next morning, and Neil Sloane [an eminent mathematician] met me at the airport because we had some project that we were going to work on. I said, "Drop all that, this idea of Richard Parker's is absolutely fantastic." I lived off that paper for the next five years.

Parker was like that. One in a hundred of his ideas was OK, and there were ten a day at least!

This type of mathematician—enthusiastic, spotting shadows everywhere, occasionally putting better calculators or theoreticians on the right scent—has been vital to the history of Group Theory, and in particular the Monster.

Simon is different [says Conway]. He has this tremendous understanding, and he cannot rest if there's some possible falsehood or contradiction lying around. Even if it's somebody on a train saying something trivial. He's like a dog gnawing at a bone. He *has* to find out what the truth is. I mean, I sort of know whether something I'm doing is deep and subtle and valuable mathematics or whether it's just a frivolity, but I refuse to be intimidated by it now. I'm prepared to do whatever I like. But Simon to some extent doesn't know. He really regards all things as equal.

Conway's mathematical strength is in his flamboyant irreverence. He discovered:

1. The Game of Life. A game in which simple patterns made on a checkerboard "evolve" according to three extremely simple rules and (in certain variants) explain the coordinated movements of flocks of starlings, expose the organizing principles of termites building up woodchips, and reveal the rationale behind the behavior of human crowds.
2. An entirely new type of number called "Surreal Numbers." He discovered them while sitting at a table in a café in Cambridge, playing Chinese checkers. The way Chinese checkers progresses as a game was, he realized, a code for a previously unknown counting system.
3. The Conway Groups in the twenty-fourth dimension (though Simon's been unable to explain to me why Conway was looking for symmetries up there).

As a young man he wanted to "make some absolutely outstanding contribution and be 'Conway, who's the best mathematician in the world.'"

Well, I don't necessarily mean the best, actually, just a certain standard . . . which is easiest to describe by saying "the best."

Other mathematicians in the Cambridge faculty would come in at 11:30, spot Conway, Parker, Larissa Queen and Simon playing backgammon in the Common Room, and scowl. When these po-faced types staggered out of their offices again at 3 p.m. for a cup of tea, the four wastrels would still be playing.

There would always be a crowd of people standing round the board "kibitzing," which meant offering advice [says Larissa]. Simon was a very, very active "kibitzer," and the first time you play backgammon with seven or eight "kibitzers" it's impossible, your brain goes into a stupor and you can't play because you hear eight different suggestions. The next stage comes when you can ignore them, and in the ultimate stage you can hear everybody. But it takes a long time to arrive there, lots of practice. You have to abandon your work for a while and concentrate on the backgammon. There was a certain vocabulary, a shorthand, you know. They would say, "Come screaming out," meaning to get out of your home table. In particular, there comes a time when no matter what you throw, the outcome is determined, and then people would say, "The dice are no longer," meaning the dice are no longer needed. And if Simon said, "The dice are no longer," people took it very seriously and stopped. They'd say, "Oh well, fine, start a new game," because on the few occasions when somebody computed, Simon was never wrong.

After his second marriage broke up, Conway tried to kill himself, and spent a week in the hospital worrying about how he was going to face his students. Everyone in the faculty knew about the attempt. He knew they were all thinking about it. They knew that he knew that they knew; and so he wondered to himself, "What would Conway do? What would be a typically Conway-ish way to deal with this embarrassing moment?"

It would *never* cross Simon's mind to think like this—to care so much about what his audience thought, or to talk about himself in the third person, as though he existed both as a human body and, like numbers themselves, a Platonic Ideal.

Conway solved the problem with combative brilliance. He borrowed a T-shirt from a friend who'd climbed a notorious

Conway once invented a pair of spectacles to see in four dimensions:

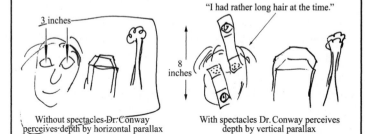

"I had rather long hair at the time."

3 inches

8 inches

Without spectacles Dr. Conway perceives depth by horizontal parallax

With spectacles Dr. Conway perceives depth by vertical parallax

Drawing by John Conway.

He used a flying helmet to hold the spectacles in place, and spent a few hours stumbling round the mulberry tree in Sidney Sussex Fellows' Garden. His brain adjusted, he walked around Cambridge with perfect ease, the first man in history to see hyperspheres, portholes to distant universes and observe the spirit world.

"No, no, Alexander. Obviously my glasses were not going to let you see an extra dimension on top of the three we can already see. That would be ridiculous. Simon told me you were not a mathematician," concluded Conway worriedly, "but you do know that we can see only in three dimensions, don't you?"

granite outcrop in California, and appeared at the lectern wearing it—the logo read "Suicide Rock."

In the Atlantis office, Conway, Curtis, Parker, Wilson and Simon kept work on the atoms of symmetry that was finished or in immediate progress in a sweaty folder that sat on a chair, growing plump, bloated, disgustingly distended; finally, obese. Occasionally, it burst.

If any of the five men had a new idea about how to solve a problem, the chair plus the folder of flab was given a forceful kick, and rattled across to him to sort it out.

Whenever it seemed the room would certainly explode with all the gathered information about symmetry, one of the men—usually Parker, sometimes Conway, never Simon—would sit down at the office's orange electric typewriter and type out the latest confirmed results on pages that were colored—with deliberate nicety—Atlantic blue.

The most troublesome Groups were draped over the tables of the Common Room outside the door. Chairs and low tables ran down the middle of this dank, rectangular space, and ended in a "refreshments" counter that was kept locked except at elevenses and mid-afternoon: 2p for a biscuit; sugar from a bowl; self-serve tea and coffee 5p a mug (mug not provided).

Anybody interested could come in, grab a cup or mug from the dirty crockery by the corner sink, splosh out a hot drink, perch over one of these sheets of crosshatched paper bearing the latest state of knowledge about the symmetries of the universe, and add their tuppenceworth. Spotted a mistake in the top right-hand-corner entries of J4? Don't think the Harada-Norton Group is quite up to scratch in the 133rd dimension? Pop in your solution, let's see . . . in this unscribbled-on scrap of margin here. See if Simon approves your suggestion when he gets back tomorrow from his bus trips to Bean (Kent) and Leek (Staffordshire).

This Common Room was the focal point of Cambridge mathematical life and, like a hospital, was open twenty-four hours a day. Even at 4 a.m. it was obscurely busy: slippered noises of people pacing back and forth in the surrounding offices; bursts of typewriter clatter; the hum of electrical fittings; a kettle boiling; the clack of dice and backgammon counters; a cry!; an office chair flung back; equations refusing

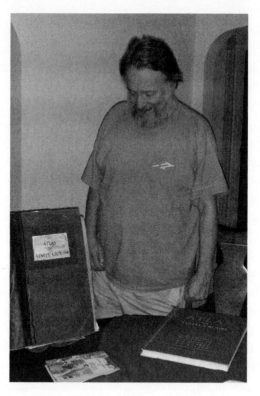

Left to right: obese folder, Conway, *Atlas*.

to cooperate; shouts, swearing, feet pounding down concrete steps to street level.

The squeak, swing, squeak of the front door and a small returning rush of fresh air.

Simon's time of Great Silence from the biographer's point of view was his time of great noise in the hyperdimensional universe of mathematics.

Simon worked on half a dozen other projects during this period—not just the *Atlas:* surreal numbers, the Game of Life, Y monograms, the biMonster. He "masterminded" the existence proof of the Harada-Norton Group.

But it is his work on the *Atlas* that keeps his name alive.

After fifteen years of constipation—

("I think pregnancy is a better metaphor," mumbles Simon.)

—Simon's most famous joint mathematical publication at Cambridge, the *Atlas of Finite Groups,* was excreted.

("Born," hurries in Simon. "In 1985," he adds.)

The mathematical equivalent of a global malaria-eradication program (the counterpart to malaria being "intellectual despair about Symmetry Groups"), the full title of the publication is the *Atlas of Finite Groups: Maximal Subgroups and Ordinary Characters for Simple Groups.* "Groups" because the book is investigating symmetries,

similar to ⬜ and △ ; "Subgroups" because that's always the way forward in this subject; "Finite" because the associated Group Tables of these symmetries have a countable number of entries; "Atlas" because it attempts to cover all the types of foundational finite symmetry known to exist, and in a certain sense the tables can be regarded as a type of map; and "Simple" because a) these Groups are all atoms of symmetry and b) the contributors are lying through their teeth.

The book is imbecilic with complexity—and littered with errors. Even with five of the best mathematicians in the university (two of them, Conway and Simon, among the best in the world) working for a decade and a half, there was too much information to master. Much of Simon's work since this date has been to do with correcting the mistakes and filling out the gaps of understanding that still remain in the *Atlas.* The introduction to the *Atlas,* written by Conway, includes the following salaam to his brilliance:

Simon Norton constructed the tables for a large number of extensions, including some particularly complicated ones. He has throughout acted as "troubleshooter"—any difficult problem was automatically referred to him in the confident expectation that it would speedily be solved.

Simon (after a haircut), Rob Wilson (another of the editors) and Conway at an *Atlas* reunion meeting in the King's Street Run pub, Cambridge. The event was organized by Siobhan Roberts, Conway's biographer. "There are so many biographers here," grumbled Conway, "we can hardly move."

As the case of Triangle and three-pieces-of-garbage-kicked around-Simon's-Excavation showed, symmetries can be categorized into types. Those two symmetries (which appeared at first to have nothing to do with each other) turned out to be identical. The job of the *Atlas of Finite Groups* was to catalogue all the fundamental types of symmetry. That's why they're called the Simple Groups. They are the atoms of the subject. There can be other symmetries apart from these, but they will always be composed of one or more of the basic forms found in the *Atlas,* in the same way that a molecule is always composed of atoms from the periodic table. The Simple Groups catalogued in the *Atlas* are the elements, atoms or building blocks of Finite Group Theory. Every Finite Group can be broken down to a collection of one of these fundamentals, just as any molecule can be broken down into a combination of the elements in the periodic table, or any building can be fragmented into doors, windows, bricks, electrical wires, bent pipes and plaster dust. The *Atlas* is therefore the periodic table or builders' catalogue for Group Theory.

This classification of Groups into their simplest components was a project of Victorian grandeur: a taxonomy (to use a third metaphor) of all the insects of Finite Symmetry—every mahogany drawer in the entomology halls of mathematics flung open, every tiny beetle sent in from a vicar's garden plucked up and investigated to the last follicle.

The largest "sporadic" atom in the *Atlas*—the biggest specialist brick, the most gargantuan and armor-plated exotic insect—is the Monster. It is the largest Finite Simple Group in the universe. There is nothing beyond it, except silence.

As soon as possible after the manuscript was published, Conway emigrated to a professorship in America, desperate never to look at a Group again.

The mathematician Benedict Gross has said that if his house was on fire, the one thing he'd battle back through the roaring flames and crashing timbers to rescue, leaping across cavernous molten stairwells, dodging the exploding gas boiler—would be the *Atlas*.

"Which is quite unnecessary," points out Conway, "because you can get it online."

Simon's copy of this immortal book—torn to halfway down the ring binding, three blobs of biryani and Ferns' brinjal pickle staining the top right-hand corner—has moved from the place where we first met it, in Chapter 4. Simon consults the *Atlas* frequently. Now, it rests on top of the twizzle-legged table, under a love letter to his mother and a new paperback murder mystery called *Zombies of the Gene Pool.*

And abruptly we are here: the critical moment, the biographical climax of Simon's story.

It's so important to get this right—to get the pacing correct, to understand its immediacy, its horror.

I'll use the present tense. It's a moment of resonance for everyone, not just geniuses.

We're back in the Atlantis office—or perhaps it was the Common Room. I'll make an autocratic biographer's decision, and call it the office. It is winter. The metal-framed windows ache with cold. Three people critical to the history of Group Theory are in the room, occasionally kicking a chair at each other between the towers of paper: Conway, Parker and Simon. They are discussing J4, the fourth Janko Group, page 188 of the *Atlas.*

Simon is sitting by the window. He blows his nose into a hand towel. He has a powerful interest in the Group J4. He's been chasing it down in 112 dimensions, and has almost caught it.

Conway—we'll put him at the desk with the orange typewriter, one hand delicately positioning the page, the other stabbing at the keys. With a Group Table that's 86,775,571,046,077,562,880 columns wide, even the shorthand versions of the fourth Janko Group flop across two pages and require supreme secretarial skills.

"Simon," says Conway, turning round, "if $2^{(1+12)}.3.(M_{22}.2)$ is an involution centralizer of J4, what is . . .?"

Today, no one can remember what the question was— something "comparatively trivial," recollects Parker.

Simon answered Conway instantly, of course.

Conway, satisfied, returned to his typing . . . then he stopped and gasped—Parker uses that word, "gasped"—and turned back to Simon. The unbelievable had happened.

"No, you've got that wrong. You must have made a mistake."

Simon spotted the error too, and blushed.

"That," said Conway, "is the beginning of the end."

And it was.

33

We have a joke in Russia: Einstein has died and gone to heaven, where he meets God. And God says to him, "You, Einstein, you have worked hard, you deserve some reward. Ask me any question about anything you like, and I will give you the answer." So Einstein thinks a bit, then says, "What is the equation of it all?" "Ah," says God, picking up a piece of chalk, "it is like this," and he writes a long formula on the board. Einstein looks at this very carefully, nodding and appreciating, then suddenly frowns and points at the formula. "But you've made a mistake."

"I know," says God.

Told to the author by a Russian mathematician, in Montreal

Twenty-five years after the publication of the *Atlas*, Simon appears in the back of a beaten-up coach, teeth bared in excitement, in the godforsaken seaside town of Scrabster, north Scotland. He has come for a holiday. His hair is bedlam—as if someone has slapped glue on his scalp and bounced him across a cornfield. Muttonchop whiskers scrabble smokily up his jaw. His eyebrows march across his forehead, one after the other, like African caterpillars.

Summer and winter, Simon wears the same clothes: a red or blue T-shirt that has started to decay along the seams, and a dark-blue puffa jacket under which he alternately sweats and shivers.

From Scrabster, we plan to take the ferry to Norway, a six-day cruise up the west coast to Lapland and the Barents Sea, then visit Russia (Murmansk, St. Petersburg) and come back to England through Belarus and East Germany. A seventeen-day journey in a heat wave for which Simon has brought two spare pairs of socks and one T-shirt.

I'm the only person in the world he knows who will be delighted to go with him.

I'm not living in Simon's house these days. I've moved to London, and we haven't met in months. I've discovered it's easier in biographies of living people to pretend your subject doesn't exist. Booted out of the door, he settles down in comfort on the page. But this much separation has started to make my ideas about Simon bland and easy, like fiction. His unsummarizable character has become simplified in my mind, even cartoonish, hemmed in by writing style and prejudices.

As Simon gets off the bus I'm quite surprised to see that he has legs. I'm taken aback that he knows how to use them to walk upright and negotiate obstacles. We settle down at a waterfront restaurant. It comes as a considerable shock to hear him order a meal of grilled sea bass and potatoes dauphinoise in faultless English.

The northernmost port in mainland Scotland, Scrabster is exactly as it sounds. A few houses and a dank bar quickly peter out into slabs of parking concrete, groups of cyclists with windcheaters and maps in plastic spongebags, who've arrived six hours too early, and a long concrete buttress against the savagery of the North Sea. Seals swim about in the harbor, nagging the fishing boats when they putter back home with the day's catch.

Across the bay is the town of Thurso: burned-out payphones and yellowing takeaways.

There is one thing only to do in Scrabster: leave.

"Oi, fucking mind where you're going! You want to keep him on a leash, mate."

Bergen, the next morning. All the way up the funicular to the city park, Simon had clutched his map, but as soon as the doors of the crowded carriage opened, a breeze dashed into the folds and puffed the huge sheet of paper out to full size:

Oslo slapped a woman across her cheek; Trondheim decapitated an ice cream; an expanse of North Sea leapt up and assaulted a man setting up his tripod to take pictures of the bay.

"On a leash, I'm fucking telling you!" the camera man raged.

"Oh, aggh, hnnnn, oh dear!" protested Simon, trying to shake the swollen map back into shape. "Hnnn, aaah, oh dear, ohhh!"

Simon is still doing excellent mathematics—occasionally. He may have vanished from the mathematical scene in some people's eyes (at least, in terms of giving lectures and appearing at research seminars), but he's still active in the remaining sixty trillion square meters of inhabitable earth outside the Cambridge Mathematics Faculty. New ideas constantly demand his attention. The one that leapt out of his sock drawer last winter is proving extremely fruitful.

It was as he was trying to decide what to put on his feet one morning that the question occurred to him: imagine that his socks, instead of all being slug-gray, came in three colors—say, red, blue and green. If exactly half the time he pulled two socks out of the drawer they were the same color (red and red, for example), what would this say about how many socks he has of each color?

It's the sort of problem that darns a grin to Simon's face for months.

He's going to write it up for a book to be given to children competing in the Mathematics Olympiads. He wants it to be called "From Sex to Quadratic Forms."

"Why sex?"

"Why not?" He chuckles.

"Why not? Simon! Because, for God's sake, it's . . . ah," I stop, remembering myself. "OK. Why not call it 'From Roast Chickens to Quadratic Forms'?"

"Because roast chickens are female."

Simon is still studying the Monster, but it is slow going. He is working his way round its Group Table, gathering up clues in much the same way that he studies bus and train timetables, picking out unexpected connections, noting tidy circular journeys of numbers, bashing off emails of complaint to the editors of the website edition of the *Atlas* when he uncovers errors in the Monster's tabulation . . . waiting for inspiration. For public transport, Simon uses the word "artistic" to describe a satisfying jaunt—one in which all his timetabling comes together with minute-perfect efficiency and he ends up back at Cambridge bus terminus at around 10 p.m., footsore, exhausted, duffel bulging with fresh tourist leaflets, and just in time to pick up a chicken biryani from the takeaway on Mitcham's Corner.

For mathematics, he prefers the terms "elegance" and "beauty" to describe such exemplary performance.

Many people insist that Simon is barking up the wrong tree. The mathematician Richard Borcherds has come up with a brilliant explanation that links the Monster to certain symmetries in String Theory—what more do we need to know?

But to Simon's mind Borcherds's extraordinary ideas are partial and inelegant, the mathematical equivalent of a cobbled-together shortcut across rough terrain in the back of a borrowed Land Rover. And since, Simon says, in Group Theory "beauty" is what guides him toward truth, the true answer is still to be discovered. All Simon needs is a breakthrough, a clue that will bind all the loose ends together; it could happen any day now. He has had so many false starts . . .

"Socks are a generalization of sex," Simon declares, swerving my thoughts back to why he's become a mathematical pornographer. "Let us suppose, instead of three colors of footwear, red, blue and green, we make the problem simpler and consider sex: let there be α boys and β girls, such that if the chance of any two selected being of the same sex is precisely half, then the number of pairs of children is given by the equation $\frac{1}{2}(\alpha + \beta)(\alpha + \beta - 1) \ldots$"

My brain spins.

The sense of panic I know so well takes over.

Starting in at the third or fourth rib below the heart, it spreads congealment and self-loathing.

Here I am, a man with a first-class degree in physics, and an MSc in applied mathematics, but my eyes tighten to my nose as soon as he begins this dreadful theoretical stuff; my ears ring, my body cramps with refusal to understand, and I'm as bad as those absurd people who giggle about how "appalling" they were at math in school, how they had to be kicked out of the class and failed their GCSE fourteen times.

Yuck.

"North! North! The path goes north here!"

Beside a lake in Bergen city park Simon suddenly turned off the perfectly well-marked pavement, into a bush, came out the other side coated in leaves, and barged up a slope. "North! Keep to the map! North!"

He held the map up high, over his head, and jabbed at an edge.

"But Simon, we were *on* the path. This rut is an animal track, for *very small* animals."

"North, north!"

I ran after him and dived in among the spruces. The sharp, bare branches were blackened with lichen and waterstains. Felled trees lanced the ground. Wood spikes stabbed at our backs and trouser legs.

Simon is an excellent guide on a walk. He spends so much time staring at his foul bed-sized maps, trying to figure out where on earth he might be, that he loses track of where he is. He takes a wrong turn, doesn't notice that he's just walked past a no-entry sign with a man holding a gun, and the result is adventures.

Plastic bags protruding from his duffel, he clumped through a bank of ferns, jolted right across fifty feet of sodden grass, and disappeared into a prickly mass of felled wood.

Simon doesn't *consult* his map: he clings to it. It is his motor. If he let go, it would zip off, veering and slewing, refracting among the branches, racing out above the mountain canopy until it hit the horizon and disappeared in a blink. Then where would he be? Half dead, in his opinion: a heap of human lostness on the woodland floor, being nibbled at by ants.

"Yes, here it is! The path! The path! It goes west!" cried Simon, weaving through this chaotic thicket. Across the clouds in the west, a bolt of lightning cracked. I ducked a ghostly branch, then instantly had to change direction and leap over a fallen trunk. I felt like a weaver's shuttle.

We emerged in desolate scrubland. For an hour after, the animal track meandered vertically, fit only for goats.

"Sometimes, when I am walking I get involved in a problem," Simon flung back in breathless explanation, "and I have to know the answer, immediately!"

"Is that what this is about? A problem?" I called ahead. "You've had a breakthrough?"

"Yes!"

These sudden bursts of insight are a famous feature of mathematical discovery. They reach out of the sky and snatch mathematicians by the whiskers.

"You've discovered the answer?"

"Yes!"

In the tumult of discovery, they react in different ways.

J. J. Sylvester, the Victorian poet who invented matrices, describes discovering a new theory "with a decanter of port wine to sustain nature's flagging energies, at the usual cost of racking thought—a brain on fire, and feet feeling, or feeling-less, as if plunged in an icy pail." Sophus Lie, founder of Infinite Group Theory, walked naked through the countryside around Paris during the Franco-Prussian war in 1870, stuffing his mathematical writings into his knapsack. Henri Poincaré, the great French polymath, once worked on a problem for weeks, made not the slightest advance, and threw it aside in disgust. Several days later, as he was getting onto a bus, the solution suddenly popped into his head. The short period of calm had been essential after all those weeks of concentration. The solution had needed time alone in his brain, as if to brush its hair and oil its mustachio. Now it was ready to kick up its heels in public. Simon Norton responds to his mathematical discoveries by jolting off the public path and rushing through woodlands in Norway. I realized: this is what Simon's public-transport jaunts are about—travel gives him ideas. It is his laboratory of thought. Simon really is *physically* hunting the Monster.

All I must do is make sure I'm the first person he contacts when the discovery occurs.

"Aagggh, uuugh . . . As I say," Simon interrupted. "I wasn't talking about a problem in mathematics. I meant I had to know the answer to where this path goes."

It is not just that Simon's perspicacity collapsed but that it vanished so drearily. From the greatest mathematical prodigy Cambridge had seen—the greatest "native" talent in the country for perhaps a century—he sank to chasing footpaths and hoarding bus catalogues. He became a cursed figure. Never, said mathematicians, had they seen such a spectacular and thorough demotion. From blessing to damnation with Classical Greek rapidity. Never a loss so tragic and complete. Unemployed, unemployable, Simon dried up like old pastry in 1985, and has been a bag of crumbs ever since. He is a morality fable about the dangers of rampant genius.

A great deal is written about genius, what it is, how it shows up, how the rest of us can snatch a slice of it, from "Eat baked tomatoes" to "Work like a crazed person for 10,000 hours." There is nothing on why it disappears.

In 2007, before this trip to Norway, I went with Simon to a conference on the Monster at the Centre de Recherches Mathématiques in Montreal. Many of the major names were there: Conway, Harada (who co-discovered the Harada-Norton Group with Simon), John McKay . . .

When it was Simon's turn to give his talk, he barged up to the front of the lecture theater as though he hadn't the faintest idea what he was going to do next. For a while he poked about in the chalk box. Was there a nice boiled sweet in there? Then he walked across to the far right of the blackboard: "Uuuuh, if the letter C stands for the language

310

Cherokee . . ." he said, and wrote the word out in large letters, with a capital C.

C-h-e-r-o-k-e-e

Good God.

"And R represents Romanian . . ."

R-o-m-a-n-i-a-n, he spelled out directly beneath the first word.

The audience gave a nervous titter.

". . . and M is . . . uuugh . . . Maori."

M-a-o-r-i.

"Then you will see that the initial letters of . . . uuuh . . . these three languages correspond to the acronym for this institute: **C**entre de **R**echerches **M**athématiques."

Most of the time, Simon does not face his audience. "I thought of that on the flight over," he told the blackboard.

Another awkward flutter passed through the auditorium.

Earlier that morning, over coffee, a young man from Texas had pushed his way through the crowd to Simon.

"You must be the man of the day," he'd said.

"Why?"

"Because you look even more like a lunatic than John Conway."

Simon had laughed uproariously—then abruptly walked away.

As he stood in front of the blackboard, sticking the chalk through his whiskers, Simon seemed genuinely intrigued by the oddity of whatever point he was trying to make, then recollected himself.

"In many ways the evolution of a language parallels the development of the Monster. In the same way that Cherokee, Romanian and Maori are seemingly completely unconnected . . ." As he spoke, Simon rubbed out the words and replaced them with:

Congruence Groups
("You'll notice I'm preserving the initials.")
Replication
Monster.

And at last, the audience began to sit up; they realized he was not mad. Pieces of notepaper were extracted from files and flattened, pens picked up in preparation. Congruence Groups, Replication and the Monster: three areas of mathematical study that had once been considered foreign to each other. The simile was as clear as the Canadian mountain air. Spotting coincidences (this was Simon's point) has been our most effective weapon in the hunt for the Monster. We were back in the territory of the sane.

As we came out of the lecture hall following Simon's talk, an eminent Professor of Group Theory from France sighed with pleasure. "Ah! *Zat* has made it worthwhile. Conway and then Norton. It eez worth flying 6,000 miles to listen to genius."

Simon's explanation for his loss of talent is that it is nonsense. He is as good at mathematics now as he ever was—better, perhaps.

In late afternoon above Bergen, we reached the top of the mountain.

"Now we want to go southwest," declared Simon. "Let's see, the sun is . . . uuggh, wait a minute, where's the sun gone?"

"That bright yellow blob there, Simon, next to the cloud."

"Ah, yes. Let's see now, right, that means west is . . . there, and north there, and so Bergen is *that* way."

"No, Simon, Bergen is that big collection of buildings down there, behind you."

"Aaah, hhnn, you might be correct, but, aaahhh, I'm not 100 percent convinced. I am aaaah, 80 percent convinced . . ."

On the descent, Simon took another wrong turn because of his map, and sent us sliding down a muddy ravine on our bottoms.

"The place we're heading for is where we were!" he cried. "I think it is very important," concluded Simon as we burst out of the undergrowth in front of an alarmed Norwegian lady pushing a pram, "that people be taught the skill of getting around."

"Uugh, hnnnn . . . Are you allowed to eat tinned corn uncooked?"

Back at our Bergen apartment, Simon had agreed to make supper—Mackerel Norton, the only dish he knows how to cook—and had instantly run into complications.

The sweet corn had been packaged in Norway.

"And . . . aaagh . . . I can't read Norwegian."

"Simon, you've been eating uncooked sweet corn from tins for two decades. It's the same food whatever language you eat it in."

"Ugggh, hunnnh . . . I'm not certain how you can know that. What evidence do you have? Perhaps . . . aaggh . . . perhaps . . . oh dear!"

Simon lives in constant disquiet about outside chances. If a thing is possible, no matter how unlikely (one plus one might not equal two; Cambridge housing officials are going to evict him as soon as they've read this book; the number in the thirty-fourth column of the 192nd row of the Monster's Group Table might be a seven; a can of sweet corn is different in Norwegian), he *has* to give it space to be heard and fretted over.

Simon approached the microwave sniffing, crouched, his brow knitted. When his nose arrived at a suitable distance from its door, he began to move from side to side, trying to see past the grid into the middle of the machine, smoothed his hand along the top, then gave one of the buttons a sharp jab.

It pinged.

Simon leaped back and scuttled out of sight.

"I am a worrier. My mother was a worrier," he agreed from the shadows.

Next, he couldn't find a can opener. A series of bangs, as of a tin being chivvied with a carving knife, exploded out of the kitchen into the living room.

I sipped my wine, leaned back on the sofa and watched the sun sink toward the horizon.

Does this, in part, also explain his "loss" of genius? He gets caught up in silly intellectual tangles, problems about socks, trivial corrections to errors in the Monster Group Table, questions concerning the way numbers fit together in the bus timetables of West Yorkshire, because, as Conway says, he has no more judgment about what's important in math than he does about sweet corn? Technically and creatively, Simon is superb, world-class—even now, despite his errors. It's his sense of direction that's gone skewy.

"Yes, that is possible," Simon replied from the kitchen in an unconcerned voice. The banging noises had stopped. They'd been replaced by clinks and thumps. He'd found something on the wall that he thought might be an electric can opener, and was trying to get the sweet corn tin squashed into it.

After Conway left for America, Simon couldn't work for a year. He was shattered, appalled, rudderless. Simon had looked up to Conway "almost as a father"—although it's not clear to me exactly what either of them means by that comment, since you'd be hard pressed to find two men with less sense of what fathers do. When Conway went to Princeton, he left Simon defenseless. He did not offer to take Simon to America with him. And with Conway abruptly gone, Group Theory work on the *Atlas* finished and packed away and no one to goad and tease him into attacking fresh problems, Simon was lost. The popular image of a brilliant mathematician is a man who looks like Simon, and spends twenty-three hours a day alone in his mother's attic solving the most difficult problem in existence.

But Simon is a different and much more common type of mathematician. For his genius to flourish, he needs liveliness and company. Isolation in a garret destroys his force. Untalkative, withdrawn, ham-fisted, a grunter when talking to his biographer, as a genius Simon needs comradeship, hilarity and someone who'll make him dance all over mathematics in a salsa.

With Conway gone, Simon had no champions and few mathematical friends. There was no one to work with, so he did not work. The Mathematics Department sacked . . .

"No!" shouted Simon from the kitchen.

. . . refused to renew his contract. The professional career of one of the great mathematical prodigies of the twentieth century was over.

"Aaah, hnnnh . . . I think I've done something wrong"—scrape, scrape, scrape.

Simon had managed at the same time to undercook and to burn the Chinese-flavor packet rice. It looked as though a moose had squatted on the plates. He gave me the spoon, with the blackened deposits still impacted onto it, as my cutlery.

"I think the other thing you have not considered," he said without sadness, investigating his first forkful of Mackerel Norton as though that, in the end, was all that mattered, "is that I was not a genius."

Fretful about the corn, he had left it in the tin. Now he spooned out a little pile of yellow blobs, took a nibble, and urgently stirred the rest into the steaming, carbonized mush on his plate. "It will get reasonably warm by the time it reaches the mouth," he suggested, with hope. Then he upended a paprika jar he'd found in the kitchen cupboard. "Shall we add some spices?"

What makes Simon doubt his genius is not modesty. Simon has no self-curiosity and therefore no desire to strut or underplay his talent: to Simon, Simon is a collection of disparate facts and no interpretative glue.

But he is a purist about language.

Einstein was a genius; the German mathematicians Carl Friedrich Gauss and Leonhard Euler were geniuses; Feynman was a genius. Not even Simon's most fervent mathematical fans would put him in their league.

But by the same argument, I point out, you could say that Ben Nevis is not a mountain, because Mount Everest is.

However, just last week, I got a text from a friend who's a secretary in the Cambridge Mathematics Department. She'd asked one of the lecturers to look up Simon's degree result. The rumor that Simon ranked at the top of his year as an undergraduate, scoring fifty alphas, when all you need to get a First Class degree is twelve—it's rubbish. I've got it wrong.

("That sounds in keeping. I have not had any ill effects from the sweet corn yet. Have you?")

Simon did not even do particularly well: a middle-rank First—maybe, thirteen alphas.

("It's not surprising that I forgot about being the best ever if it never happened, is it?")

By no stretch of the imagination (going strictly by exam results) can it be said he was the most brilliant undergraduate mathematician Cambridge has had since Newton. Not even the best of his year.

Is Simon therefore now correct about his lack of genius? Did he lose his ability fifteen years earlier than this biographer thought?

No.

At Cambridge, if you want to do research, you have to take a one-year postgraduate cramming course known as Part III. It's a harsh test, designed to ruin relationships with girlfriends, crush illusions, destroy the green shoots of manly pride and lead to early-onset alcoholism. I took it, and the less said about that the better, yet this is the only occasion on which Simon and I have been mathematically comparable.

("I think *that* is an exaggeration!")

He also almost failed. Tracts of questions left unanswered.

The lecturers shook their beards out of their sherry at the shock of it.

Had Simon been an ordinary student, this would have been the end of his career in the faculty. It would have slowed down the research to have such a bumbler about.

Fiercely, the lecturers bubbled at their tobacco pipes. It was inconceivable that he could have found the mathematics difficult. There was nothing for it but to amble up and down the Senior Common Rooms arguing violently in Latin and Greek, then to let the peculiar boy stay.

The explanations for Simon's average First and his awful Part III are the same: he was bored. It's all very well saying that what you need is 10,000 hours of study to become a genius at a subject, but the genius is not in the hours, it's in what makes you want to do such a foul amount of study: unless it's delight, you won't get a genius at the end of that time, you'll need to get a shovel to scoop up a suicide.

Eton was good for Simon, but they made one bad mistake: they encouraged him to start a university degree when he was fifteen. He did brilliantly. A top First. Extra prizes sprinkled on top. Name splashed in the national press.

But when Simon went to Cambridge, the lecturers there—possibly out of sheer arrogance—encouraged him to take the second half of his undergraduate degree all over again. For a whole year, at a time when his brain was at its peak of receptiveness and joy, they tortured him with repetition. Looked at in this light, he's lucky his genius didn't collapse altogether.

That's what happened to his ex-Olympiad partner, Nick Wedd, last seen breaking into Simon's rooms in Chapter 29.

Like Simon, Nick had had a phenomenal intuitive gift. By the time he was twelve, he could write down answers in math lessons without any calculation at all (although he'd learned that "I had to provide workings, which was a string of gibberish that kept the teacher happy. If there was enough gibberish and then the right answer then you got a tick"). At the Mathematics Olympiad in 1969, in which Simon achieved 100 percent for the second time, he scored the second highest possible award. Many postgraduates couldn't do nearly as well, and Nick was fifteen.

The only thing that can explain what happened next is that his schoolteacher went mad. When Nick returned in triumph, instead of celebrating his talent and allowing him to indulge it, this absurd man forced him repeatedly, for two years, to sit in the schoolroom and solve trivial A-Level problems. PET studies measuring the brain activity of calculating prodigies have found they recruit different regions of the brain from ordinary people. They don't perform their sums in the same dull and schoolish way as the rest of us. But this teacher was determined, once and for all, to bring Nick down to a level of pedantry he could understand and for him to "provide workings." The result: Nick's ability ruined by boredom. One of the finest young minds in the country two years before, he got a C in his A-Level exam and was told by his school not to bother applying to study mathematics at university, nor ever to think again about being a mathematician.

The insistence that Simon redo his final undergraduate year at Cambridge even after gaining a superb degree in London meant that he skipped his undergraduate lessons and wandered down the corridor to the more interesting postgraduate, Part III lectures. Then, after he had yawned through his finals for the second time, the university administrators insisted he do the Part III *officially,* even though he'd already *unofficially* taken all the courses that interested him—his second year, therefore, of tedious repetition. Simon went to sleep.

It was only once he got past this silly administrative process and discovered Professor Conway that his genius again found its narrow but supreme strength.

To prodigies, talent doesn't come from hours of hard work, it comes from delight. As long as they find what they do delightful, they'll keep at it. But over-ambitious parents, inflexible math teachers and humdrum university programs can destroy the delight in as little as six months; shortly after, the brilliance withers away too. Conway believes it is almost always a bad idea to send math prodigies to university at an early age. They rarely, when you do, come to anything as adults. It is too early to destroy a child's social life and regiment his thought. Let the child roam. Give him an expert tutor, but for as long as possible let him stay free and guided by delight.

"As I say," said Simon, finishing off his plate of mush and the last speck of sweet corn with a sigh of satisfaction and leaning back to watch the sun disappear into the sea, "I call the time Conway left Cambridge my bereavement."

34

Me: Looking back, if you could change one thing about your life, what would it be?

Simon: I would have been more sensible. I would have taken more care over the routes I took.

Me: You mean you'd have taken more care in the type of work you did for your mathematical career?

Simon: No. I mean I would have taken more care planning my bus routes.

Aside from a recent mugging on Jesus Green in Cambridge, when three men forced him onto his knees and made him beg for mercy because he looked like a homeless person, Simon's life has not been troubled by excitement during these last months.

He doesn't look on time as most of the rest of us do, moving from incident to incident; he dwells either in satisfaction or out of it. You can bob along the west coast of Norway with Simon, staring for hours at surging mountains and fantastical contortions of rock face, without feeling that a single thought has passed through his head.

His sole phrase for everything—from the vertiginous cliffs and crash of mountains to the langoustine, crevettes and lumpfish caviar that cruise boats lay out on platters like pearled swatches every lunchtime, is "It's alright."

"On your right, you see the town of Standa!" shouted the polite ferry guide on our excursion down Geirangerfjord, after our day in the Bergen woods. "Standa has the only pizza-making factory in Norway, thank *you.*"

Palisades of cliffs thundered against the sky; waterfalls— braided, crashing, bounced out from precipices, exploding diamond speckles high across our heads—plunged into a muffle of trees a thousand feet above, and re-emerged as bathtub bubbles next to the hull of our boat. Simon's response? "It's alright."

"And please don't spell it 'alright'. I have always insisted strongly on spelling it as two words, so please do the same."

He feels certain spellings are wrong in a "visceral" sort of way.

"*Ja*," continues Polite Ferry Guide, "and on those farms on the cliffs, livestock and children must be kept tethered to stop them falling over the edge—thank *you*." Toppling crags and looming escarpments blocked out and released the heat of the sun. A sea eagle soared in predatory arcs among wisps of cloud. "*Und* the only way to get a cow there is to carry it up as a calf. *Ja*, in the nineteenth century, when the taxman is visiting, the farmers hide all the ladders which stop you falling off the mountain."

The water in the middle of the fjord, still as ice, reflected clouds. It seemed as though the mountains were flinging the sky itself back and forth between their peaks.

"It's all right," asserted Simon.

"Can't you think of *any*thing to say but 'all right'?" I cried.

"My mother used to complain of the same thing," agreed Simon contentedly.

As a matter of fact, I'm not unsympathetic. How do you describe the west coast of Norway? What words could possibly take on the job? At the end of this fjord is a tourist center, where another cruise ship was berthed next to the knitted-goods shop, its house-sized propellers murmuring; hundreds of tourists were on the dock, each carrying a shopping bag containing a squash-featured, pot-bellied, turnip-limbed doll with floor-length pink nylon hair.

Mathematicians, in order to make progress with the notion of infinity, often talk about it in terms of division by zero: zero will go into seven (for example) a limitless (i.e., infinite) number of times. To rephrase infinity like this brings the idea closer to us, and a little further from mysticism. It hems the subject in a little. Everybody can relate to zero. Norwegians do the same thing with beauty: faced with the inexpressible splendor of their mountains and the glacial stillness of their fjords, they have invented trolls.

The journey up the coast of Norway into twenty-four-hour daylight takes five days and is accomplished by the eleven liveried ships of the Hurtigruten line, which are in constant rotation, bringing supplies and vast parcels on pallets to the coastal villages. They greet each other among the islands with a clangor of sirens. Our ship was the 11,000-tonne *Richard With,* named after the man who set up the company around the turn of the last century.

Each evening I pushed Simon's unbending socks off the side table, set up my digital voice recorder and mic, and tried to conduct an interview. But it was hard to fix his attention. Every time a mountain passed by Simon's nose rose furtively, drawing his head and body up after it—as if he really thought I might not notice—so that he could get his eyes over the porthole ledge and see what interesting coastal landmarks he had missed.

Or he was afflicted by a sudden starvation, and had to lunge around his duffel looking for Bombay mix, then re-emerge, sucking grease and beige bits off his fingers.

As I clipped the microphone to his T-shirt, he fidgeted and bounced and flicked through his *Thomas Cook European Railway Timetable,* wondering: "Where is Chernobyl? Is there a train going there that stops to pick up radioactive passengers?"

"Simon, please! Concentrate."

"Sorry."

Sometimes I tried a long run-up. I retreated back half a century to the horizon of his existence—the most basic questions; the necessary conditions of life. Take a deep breath, clench fists and . . . "Right, ready? Good, right, *go!*" . . . zing forward:

"When were you born?"

"Twenty-eighth of February 1952."

"What was your mother's name?"

"Helene."

"What was your father's name?"

"Richard."

"What was he like?"

"I'm sorry. I don't understand such philosophical questions."

Simon's mind can't keep off public-transport campaigning. He insists that I take this "opportunity" to include "a mention of his conversation with fellow passengers who have been inconvenienced because the link between Lerwick [Shetland] and Bergen has been axed, so they cannot enjoy this trip of a lifetime with the same ease that we do. Links from Scrabster and Newcastle have also gone."

"It is important for readers to learn at least some useful information in your book," he remarks tartly. "You can also add that I was *not* looking up services for radioactive passengers, I was working out the bare bones of how a Hurtigruten-like service could usefully be introduced in the Western and Northern Isles of Scotland, stimulating our sustainable-tourism industry. Now that I have worked it out I am wondering how to promote the idea, and suggest anyone who is interested get in touch with me via the publishers or the author's website."

"Right, good, returning to the biography . . ." I clicked on the record button. "Why do you think your genius vanished?"

"Aaaaah, hhnn . . . Can we stop now? I think I'd like a banana."

As you go farther north, the houses become increasingly blockish and made with corrugated iron. The requirements of survival are evident: oil-storage tanks, fish-processing plants, mobile-phone masts, shipyards—instead of being set aside from the town, these sorts of municipal landmarks start to gather round the high street with the houses, as if cringing against the cold. By the time we reach the North Cape the battle to jolly things up is lost.

On Meagre Island, 120 kilometers above the geographical treeline, the wind blasts away any vegetation above toe height.

"Here they have two supermarkets," announced the tour guide with a proud puff, "and, *ja,* a retirement home, thank you!"

There is more life in the air and much more underwater than on the land of Meagre Island. In winter even the reindeer vamoose. The 7,000-strong herd that grazes this island, gnawing the specks of lichen from the rocks, belongs to six Sami families. Every spring they ship the reindeer over on military landing vessels, and at the end of every summer, before the weather turns absurd with nastiness, they herd them up again, goad them to the edge of the water, and make them swim across the strait to the mainland. After that, it's a three-week trek through Norway, over Sweden, back to the Finnish–Russian border, where the Sami have their farms. On the few days of the year when the water reaches above 10°C the locals themselves nip out to a dire fifty-foot stretch of gray sand to have a swim. They call it Copacabana.

"Simon, your brother told me about David Elton, who you used to play with as a child in the summer holidays, when your family went to the coast, where you went swimming. Did you know that he became a murderer?"

326

But Simon has a knack of treating queries as if they are the end of a conversation. His brain simply substitutes a period in place of the question mark, and his face goes on smiling with no adjustment.

He sat back on the tour-boat bench with a sigh of happiness.

"With a champagne bottle?" I pestered. "His wife: he bludgeoned her over the head, then drowned himself. *Simon!*"

Simon's nose was above the porthole again. A small village rippled past, barely more than a tarnish of houses among the waterside rocks. Against one of the corrugated-iron walls, a baked moose skull, stuck on spikes.

"What does it feel like," I continued eagerly, "to have gone swimming with a murderer?"

"Aaaaah . . . I didn't like swimming."

Puffin Island is a motorway diner for birds. The sea serves up huge fish suppers all summer long to thirty-five different types of seabird that react by screaming and fighting and killing each other as they jostle for table space and turn the rocks white with excrement. The puffins are so tubby that they land on water by bumping into little waves, bouncing their plump breasts from crest to crest until they stop, and sink.

On the way back from Puffin Island, an awful, high-pitched noise came from behind my left ear. I turned to see Simon, bleating foully—"humming," he pronounced it.

"Beethoven's Sonata No. 19. A prime number!" he declared joyously.

The ecstasy of the birds had driven him to music.

Night above the Arctic Circle is not like ordinary daylight. It is autumnal light, a bit egg-yolk-colored, but also gray. Warm, cushioned, soft, it makes you feel slightly sick as you move north up the coast; but each night I tried to stay up on the boat, to make sure the sun truly didn't ever pop off for a snooze, I fell asleep before the miracle happened: somehow, between 4 a.m. and 6 a.m., the day is replenished with freshness. Strawberries and cherries grow in north Norway, dense with the sort of sweetness you normally read about only in children's books, force-fed by twenty-four hours of sun. In winter, said a banker from Tromsø, the day isn't really quite black either, but blue as if just drowned.

Once, 2 a.m., the ship hit a storm. The prow tossed up and belly-flopped into the waves. The rain slashed in under the overhanging lifeboats, then appeared suddenly to be defeated, leaving a narrow sheltered section beneath their keels, under one of which I stood on deck, feet spread against the pitching of the sea, and drank my first can of beer of the morning in a state of pleasant gloominess.

"Opening only by authorised personnel!" read a sign on a gate in the deck railing. "Danger of life!" It unbolted onto the Beowulf waves.

Three years ago, the cook on one of these ships sneaked out, dropped a small blow-up lifeboat into the sea, clambered over this sign and vanished off to the islands with one of the waitresses. I understood his sentiments. It took the police fifteen weeks to find them, radiant among the moss and puffins.

The next morning—or perhaps it was evening—it might even have been the middle of the night—there was an important bridge coming: a famous wonder of the North, soaring like a ghostly arc of paint high above the water. But Simon (standing on the deck) had lost his place on the map: he had to find where the bridge was on it before he could do anything else. For a while I watched him desperately trying to outwit the wind that rushed frigidly down the walkways of the boat and buffeted the page.

"Oh dear, oh dear!"

"Quick, Simon, quick!" I goaded. "You're going to miss it!"

A gust again snapped the map from his grip; he bounced

comically about the deck grabbing at the air to retrieve it, even as the shadow of this gloriously precise creation of bridge engineering—a mathematical astonishment—clutched on to the prow and started relentlessly to slide down the sides of the hull.

High above, on the top of the arc, I could see a small crowd gathered. We didn't wave to each other. We were just humans, protected by roughly equal masses of steel and concrete in the silly vastness of the North—one set of us suspended in the air, the other, on water—neither where humans were supposed to be.

A quarter of a mile farther on, Simon at last got his paper under control, pinpointed the spot, "Stawr . . . sigh . . . sun-det Bridge," breathed a sigh of relief and looked back.

"It's . . . *really* good!" he barked, and tittered with merriment.

Simon has two further points to make about his brilliance.

The first is that everyone is mistaken—he never was a great brain, just a very quick one. He reached the peak of his ability, lickety-split, in ten years. At five, he could do the mathematics of a twelve-year-old; at twelve, he was reading university textbooks on complex numbers; at fifteen, he was better than many university research students; by twenty, the equal of a professor, only his reading was not as broad. Then Simon's brain stopped developing. The alchemical process fizzled out. Other people began to catch up. They wouldn't reach him for a further five years, but what they call his "loss of genius" is actually their arrival. So used to thinking of Simon as miles ahead, a speck dancing on the distant mathematical horizon, they didn't know how to appreciate him when they were finally alongside, celebrating the subject as equals. They mistook equality for Simon's decline, and declared that he'd suffered a catastrophic intellectual failure.

What else, argues Simon, could explain the fact that despite his infamous "collapse" he is doing math today that is as good as, if not better than, any he has ever done before? Witness his performance in Montreal. Witness his great discovery of "the appearance of Conway Group in the projective plane presentation of the Monster" (*Simon:* "I don't think I can make it any more comprehensible to your readers than that"), done two years after Conway had left for America, long after Simon's supposed "first mistake" in the Atlantis office. Witness his paper on socks.

Simon's second explanation of his loss of mathematical direction is heartbreaking. Now that Conway has fled to America, there is no one in the mathematical world who will work with him. They say he is too peculiar, too shabby, too old. His interests are fixed in mathematics that has had its day. His brilliance is frigid. His talent, perfectly suited to an extraordinary moment in algebraic history (the symmetry work at Cambridge during the 1970s and 1980s), is out of fashion.

"I sometimes think that I would not have been capable of doing outstanding work in any field other than what I worked in. In due course I had worked out the field which I was expert in, and the cast of my mind was not amenable to diversifying."

He is, somehow, both too meticulous and too playful for modern theoreticians.

Too meticulous, because he is like a man at the airport check-in desk, madly searching through his luggage for his ticket long after everyone else has boarded the plane: Simon continues to be obsessed with exposing errors and patterns in the Group Table of the Monster, believing that the secret of the universe is hidden there, even though everyone else has taken the hunt elsewhere.

Too playful, because . . . well . . . because he is equally likely to write a paper about socks.

At 3 a.m. I left Simon on deck, in a state of vertical trance: joyous, windswept and in possession of all he saw.

Simon is not heartbroken.

At least, not about anything to do with his life in mathematics.

About the state of our public-transport services, it is a different matter:

> I'd say that you ought to treat me as if I was currently watching the great love of my life being slowly murdered, torn between my desire to save her and expose her murderers and my wish to spend as much time as I can with her while she's still alive.

35 Moonshine

With the Moonshine Conjectures, which Simon played such a large part in formulating, it was as though Simon opened the door between two very strange areas of maths but decided not to go through it.

Umar Salam

The symmetries of Triangle, and the way they interact with each other, can be written out as a Group Table with three columns and three rows:

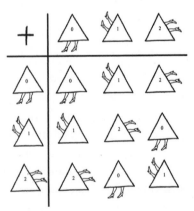

The symmetries of Square have a Group Table with four columns and four rows. The symmetries of a garbage-bag-being-kicked-about-with-three-pieces-of-rubbish-inside result in the six-by-six Garbage Bag Group Table. But what is the monster whose symmetries are documented in the Monster Group Table, with its

808,017,424,794,512,875,886,459,904,961,710,757,005,754,
368,000,000,000
columns and
808,017,424,794,512,875,886,459,904,961,710,757,005,754,
368,000,000,000 rows?

Simon doesn't know.

For reasons inexplicable to biographers, he knows that it lives in 196,883 dimensions.

But he can't explain why.

In 1979, Simon and Professor Conway discovered another thing about this mysterious monstrous object. The result was so shocking, so astounding, so contrary to what anybody expected, that they called the discovery Monstrous Moonshine.

But Simon can't explain what Monstrous Moonshine is either.

I think how it goes is like this. In 1979, John McKay, a mathematician in Canada, whose office is so messy that he once lost his baby son among the papers, discovered a remarkable coincidence. The set of numbers that explains why any object that has the Monster Group of symmetries must live in at least 196,883 dimensions cropped up in another, entirely unrelated, area of mathematics.

"Huugh, aaah, no. Alex, you've got that not quite right. What McKay discovered was that the Fourier expansion of the j-function, j(tau), where tau is the half period ratio, has as its first coefficient 196,884 . . ."

"196,884," I point out with admirable speed. "Not 196,883, then?"

"Exactly. That's the point. 196,883 + 1 = 196,884, so that shows that in 196,883 dimensions . . ."

John McKay sent Conway a postcard explaining his discovery. Fortunately for Conway, when the card arrived Simon was away on a two-week jaunt around the railways of England. By the time Simon got back, Conway had made significant progress toward explaining the astonishing coincidences, otherwise the Monstrous Moonshine Conjectures might have been entirely Simon's work.

"Hnnnnhh, no. That's not correct either. No one has explained it satisfactorily! What Conway and I did, as I say . . . aaah, let me see . . . perhaps . . . I should begin by explaining what hyperbolic geometry is. If, uuugh, I put this map on the ground to represent aaaah, huunh . . . Euclidean geometry . . ."

"Thank God I had that two-week head start on Simon," Conway is quoted as saying in *Finding Moonshine,* Marcus du Sautoy's book on the subject, "else I wouldn't have got a look-in!"

For me, the familiar panic sets in at this point, and instead of having the sense to make Simon go back over anything I don't understand, I narrow my eyes, try to look as if I'm savoring each new theoretical observation with the enjoyment of a sage, and feel the earth slip away.

"I think of myself," says Simon proudly, "as a fixer."

His campaigning letters to the local member of Parliament, badgering of government ministers about timetabling, demonstrations in Cambridge concerning the new guided-bus scheme, protest meetings about traffic-easing schemes for cars on the A14 that could be just as readily achieved by providing three wiggly buses from Cambourne plus a double-decker to Six Mile Bottom—in all these cases, Simon sees his role as that of a person who spots flaws in the way public-transport numbers and schedules have been totted up, and "brings them to the attention of the best people available to have them corrected."

"You could also say I am a hub," Simon adds. "I pass the information out along the spokes. The people who make the actual changes and see that the machine works better as a consequence of what I have shown them are the rim of the wheel. What you call my jaunts are part of this campaigning, because how else can I report on what is going on, except by experiencing it myself?"

"And with the Monster and Monstrous Moonshine . . .? Is that what you think of yourself doing there too? You are still going over all the facts and calculations involved, spotting errors in the Group Table and reporting them to other mathematicians, being the fixer? You have, so to speak, not stepped through the door and got on with further profound mathematics, because you got distracted by the door frame?"

"Yes, aaah, huunnh, I suppose so. As I say, I never understand you when you are philosophical. I don't think you make sense. Incidentally, when I went to buy a new bag and hiking boots, the shop I usually use was a hole in the ground. But I have not thrown out the old pair. I'm keeping them to hand, in case a nail comes up."

"Thank you, Simon."

Sitting in our cabin one night on our trip to Lapland, I tried another approach to get the subject back to understanding the Monster and Moonshine. I started talking about primes. Simple Groups such as the Monster—these atoms of symmetry—are often compared to primes because primes are numbers not divisible by any other number except one and themselves: primes are the "atoms" or building blocks of numbers. 2, 3, 5, 7, 11, 13 . . . are all primes. 4, 6, 8, 9, 10, 12 . . . are not, because each is divisible by other numbers. E.g., 12 can be divided by 2, 3, 4 and 6. Every number in the infinite universe of numbers can be reached by multiplying primes together, but it can also be easily shown that there are an infinite number of primes . . .

"Not necessarily," interrupted Simon. "Not if it's a pring."

"What's a pring?"

"A ring. I invented prings."

"A pring is a ring?"

Simon nodded. "With power."

"So what's a ring?"

"Uuugh, let's see . . . if you have an, uugh . . ." he began, with the same bright-eyed optimism with which he starts all his attempts at mathematical explanation with me, although for some reason the hesitancy had come back. "No, don't interrupt, the quickest way to make progress with this is for you not to talk . . . if you have a mapping from R cross R into R, in the case of, a multiplicative and . . . no, I mean, given a binary operation on an algebraic structure that is a homomorphism . . ."

When Simon was awarded a mathematical fellowship at Cambridge it was (according to rumor) specifically added into the job description that he could have the post only if he promised never to teach.

This much I do know: Monstrous Moonshine links the Monster to distant mathematics and the structure of space in ways that are as awe-inspiring to a man like Simon as it would be to an astronaut to step out of his space machine on Jupiter, and find a Sainsbury's bag floating past. That's why it's called "Moonshine," because mathematicians can even now hardly believe it.

"I think," said Simon, standing up from his berth and shaking crumbs and clotted blobs of oil and fish off his T-shirt onto the covers, "I can explain to you what Moonshine is in one sentence."

When he really tries, Simon can be a model of clarity.

"It is," he said, "the voice of God."

Gold flakes bouncing in thick liquid are stars. Ultraviolet light makes blue shadows glow moonishly. Flashes of ginger and green are asteroids.

To ease the pain of another day of failing to understand Simon's Moonshine Conjectures, I have invented a cocktail called Monstrous Moonshine. Two swigs induce a state of contented idiocy that's similar to the look Simon has when he's back from a week on the buses.

Monstrous Moonshine contains, in whatever proportions suit the torment of your mood (the given quantities below are a starting suggestion):

Recipe for Monstrous Moonshine

2 x Absinthe
1 x Blue curaçao
3 x Goldwasser
    ("Isn't that a euphemism for urine?"—Simon)
3 x Soda water
Lime juice, mint leaves and fresh ginger
A peeled lychee
Dry ice

Mix the first seven ingredients, drop in the lychee, add dry ice (the drink will instantly start to boil with great violence) and drink in ultraviolet light through bared teeth.

Important Note for Historians of Cocktails: The absinthe is to represent moonlight—wormwood, the most important ingredient in absinthe, glows in ultraviolet light. Blue curaçao also glows in UV, and the color suggests night. Goldwasser is the oldest liqueur in the world, invented in

1598, and tastes disgusting. You'd think that after 400 years of tinkering Goldwasser would be drinkable, but it isn't. However, it contains flecks of real gold, which represent the stars. Lime juice is to offset the filthy taste of the Goldwasser.

The other trouble with Goldwasser is that the flakes don't stay afloat. They sink to the bottom. Soda water makes them bounce about.

The lychee is, of course, the moon.

Dry ice is for monstrosity. I get mine from a Professor of Chemistry. You mustn't use dry ice that comes from a nightclub machine. That's toxic. The type you want comes in small, hazy pellets and will freeze a hole through your stomach if you swallow it.

36 Discovery!

> Simon is quite likely to be the person who understands the
> Monster in the end. Why? Because Simon's Simon.
>
> *Professor John Conway*

I've seen a mathematical discovery! I witnessed it start! 6:37
a.m., at the Hurtigruten breakfast buffet counter, Deck Three.

What sparked it off?

Pickled beetroot.

Simon's like a rat, or a dog. He likes to find his base before
he sets out on forays. In new towns, he has first to race
about—bag banging wildly against his legs, practically
knocking him off his feet—to pinpoint the public library and
the biggest bus stop; only then does he relax, breathe properly
and head for the medieval castle. In a buffet, it's the same. Even
if there's not a single other human between the restaurant
entrance and the mountains lining the coast of Norway, he
wants to know precisely where and at which table he's going to
sit, and to establish a presence there by dropping his bag on
the salt-and-pepper pots. Only then will he go up to the
counter to collect his food.

I met him loading his plate with beetroot.

For Simon, a good breakfast is made up of different ways to
ingest vinegar.

"What number table are you, Simon?"

"I'm twenty-one." Then he hesitated, looked at me . . . "No,"
he corrected, "I'm not twenty-one, *I'm* fifty-five, my *table* is
twenty-one."

Provoked by grammar, he leapt to mathematics, and when I looked back at him for a moment from the grapefruit counter he was standing with his plate lifted to his lips letting *jus de beetroot* dribble down his T-shirt and pool at his feet.

He had spotted a chink in mathematics. It had to do with the shared properties of the numbers twenty-one and fifty-five, and the way sunflowers arrange their seeds and scales are arranged on a pineapple, and then something else about triangles, best thought of in terms of piles of policemen standing on each other's shoulders riding motorbikes, and how many other pairs of numbers shared these same properties . . .

Simon's greatest mathematical discoveries have often begun with frivolous musings like this. For the rest of the day I could not get a word out of him.

Could it be that he had found a whisker of the Monster?

No.
    False alarm.
    It was nothing.

37

> I don't mind you making stuff up if it *might* have happened;
> what I do mind is things that would *never* have occurred.
>
> *Simon, email to the author*

Simon agrees that the problem of trying to translate a three-dimensional person with five dimensions of taste and ten sensual dimensions into a sequence of black letters on a two-dimensional page commits you to some degree of biographical fudge, especially when the subject you're working on won't talk . . .

"He, he," titters Simon.

. . . can't remember his childhood . . .

"Heh, heh!"

. . . has no sense of anecdote, isn't interested in analyzing his abilities, his attitude to life or his relationships with other people . . .

"Huh, huh, huh!"

. . . and works on a subject of cosmic importance that no one but himself understands.

"It is not *all* my fault, you see!"

But he is not prepared to agree that in order to tell the truth about him you have to lie. To get around the flattening simplifications of nonfiction you have to be, shall we say, *creative* with facts.

On the bumpy minibus journey from Kirkenes through the no-man's-land of Karelia into Russia, I invented another method. I would manufacture our conversations before we had them: write down everything I want to ask in this future exchange, make up all the replies Simon will give, then read the result through to him. If he agrees with what I've made him say, we keep it.

If not, he has to speak or else forever be misrepresented.

English has no tense for a biography of Simon. Not "he will have done" this, but "he will did" this.

To my shock, Simon enjoys this new technique, although I'm never quite sure how much he understands it. If I send him a section of future conversation in which he will said, "I collect bus and train timetables because I am a mad mathematician with electrified hair and believe the Monster might have hidden the secret of the universe inside them," he replies, "I'm sure I didn't say that."

What he should say, of course, is "I'm sure I *will not said* it, because it is balls."

The first city you pass after leaving Norway is Nikel, the Russian center of nickel smelting. Looked at on Google Earth, it's a necrotic pucker of black beside the Ice Age hills of Lapland. The smokestacks dotted between the apartment blocks and corrugated-metal processing yards spit 100,000 tons of sulphur dioxide into the atmosphere every year—four times the total quantity from Norway. Each summer, acid rain burns holes in the residents' umbrellas.

For this section of our trip, I had pre-prepared a discussion about beauty.

Alexander: "Simon, what is beauty?"

Future-Past Simon: "Huunh."

Alexander: "'Mathematics possesses not only truth, but supreme beauty,' says Bertrand Russell. 'A beauty cold and austere, without appeal to any part of our weaker nature, without the gorgeous trappings of painting or music, yet sublimely pure, and capable of a stern perfection such as only the greatest art can show.'"

Future-Past Simon: "Aagguuh."

Alexander: "You'd think Russell was talking about icicles. G. H. Hardy, one of the greatest British mathematicians of the early nineteenth century, is equally frigid: 'The mathematician's patterns, like the painter's or the poet's, must be beautiful; the ideas, like the colours or the words, must fit together in a harmonious way. Beauty is the first test: there is no permanent place in the world for ugly mathematics.'"

("Shall I put you down for another going to said 'Uuugh' here, or would you prefer 'Huunh'?")

Simon: "Eaargh."

Alexander (waving the book he is reading, Why Beauty is Truth, *by Ian Stewart):* "Look! Here's another. Paul Erdos: 'Why are numbers beautiful? It's like asking why is Beethoven's Ninth Symphony beautiful. If you don't see why, someone can't tell you. I *know* numbers are beautiful. If they aren't beautiful, nothing is.' When it comes to describing what you do, you mathematicians are as pompous as painters. However, I take it you, like all other mathematicians, agree with these three quotes?"

(Delete as appropriate: Simon nods/~~Simon shakes his head~~/ ~~Simon has fallen asleep~~)

"So, this is the point: it's got to be rubbish. Why should maths pay any attention to whether a human thinks it's beautiful or not? If the truth of mathematics lies in its objectivity and its existence away from messy emotional humans, how can you use, as a primary test, one of the most subjective human ideas of all, i.e., beauty? It's hopeless. There must be ugly mathematics that's true, just awkward."

("Now, Simon, are you listening? This is what I want you to will said next:")

Future-Past Simon: "I think being able to see this beauty in mathematics is a sign of your talent for it, which is why this recent talk about it taking 10,000 hours to become a genius is nonsense, because you cannot learn aesthetic pleasure, only learn how better to fake it . . ."

But at this point Simon put down his map, undid his glaze and joined the conversation properly, and his answers were, naturally, much better than anything I could invent.

"It is true that there is ugly mathematics, but it is unlikely to be important. But if the result is important—for example, if it brings together areas of mathematics that were previously thought to be unrelated—and its argument is short, the argument is almost certain to be mathematically beautiful."

"Economy is a vital element of mathematical beauty?"

"But at the same time, one reason I started research work on Simple Groups such as the Monster was because they seemed to be in a fundamental way different from other Simple Groups—what you call Groups representing atoms of symmetry—not just because of size. They were called Sporadic Groups, because in a certain sense which I won't try to explain they didn't fit neatly into the way other Groups could be classified, and I thought that was ugly."

"Tidiness is also part of mathematical beauty?"

"But when I finished my PhD and hadn't managed to bring these Sporadic Groups into the regular pattern, I felt that being Sporadic was beautiful."

"In short, this mathematical beauty is *not* like other beauty: it should be mean-spirited, link parts that don't usually fit together, like stitching a bit of ear to a piece of nose, and is in general short. It would never do on a human. The only thing that does make it like ordinary beauty is that it's extremely unreliable, not a good objective measure of anything, and can change."

"All I can say," he barked, abruptly annoyed, "is that whatever nerve it was that tingled when I saw Torghatten Mountain from the boat two nights ago, it is that same nerve that tingles when I see a piece of beautiful mathematics!"

It is the best reply I have ever heard from a mathematician on this strange subject.

For the next half-hour Simon pursued facts around the index pages of his Thomas Cook timetable, snapping little glances at the map and out of the minibus window to make sure the dots and contour lines weren't getting up to mischief. Practice has trained him to read tiny print during the roughest public-transport drives without feeling sick.

Through the window, Nikel was sucked back into the bowels of the earth, a half-emitted stool, smeary with rain.

Our minibus took us to Murmansk, the most northerly ice-free port in the world. The hills behind the city docks have been chewed up to make room for banks of cigarette-packet-gray housing blocks.

The small bar at our hotel was crowded with prostitutes. Squeezed around the edge of the disco-lit room, they fidgeted and pecked at cushioned gold lamé handbags; their lips bounced and wiggled under their nostrils. Occasionally a woman would give an angry squirm, as if a breeze had got in and made a saucy attack beneath the table. Two or three businessmen sat in drinking cubicles, their jackets unbuttoned, studying their mobile phones. Britney Spears (very popular in Russia) rattled a vodka bottle on the barman's shelf.

I paid, rather stiffly, and walked out with three bottles of water—one more than I wanted.

In the foyer, Simon had dropped into a sofa among a crowd of half-dressed, buyable women, and was in a panic. There had been, as there sometimes is when Simon and I arrive at a foreign hotel, some "confusion" about whether we had booked a room there or not. While this was being sorted out in angry whispers at the front desk, Simon had emptied the contents of his duffel onto the carpet. As he scrabbled through the pile, a blot of crackling plastic shopping bags, rigid socks and finger-stained railway timetables spread across the floor among the girls' legs. Wearily, they uncrossed their thighs, dropped their knees lazily to another slant, then recrossed as soon as Simon—unexcited by these fleshy movements—had panted over to another part of his gutted luggage.

"Oh dear," he said, stretching the word out along a sigh. "Where have you been? What is happening? I can't find my gout pills."

Though short-sighted, Simon "doesn't have time" to buy glasses.

Simon, immersed in the kingdoms of the immaterial, started awake.

The 1,500 kilometers between Murmansk and St. Petersburg are 70 percent trees and 30 percent lakes, forest tracks, villages of sodden-looking huts, and sleep. The train takes twenty-eight hours to hummm, *cluck-cluck,* hummm, *cluck-cluck,* hummm, *cluck-cluck* through the Kalevala lands. Small stations get a one- or two-minute stop. If new passengers don't instantly scramble on board with their tartan laundry bags and cardboard boxes wrapped in rope, the train moves off with the clothes and packed lunches flying off behind. At larger stations, the wheezing engine stops, the brakes release a long, screechy puff and the wheels fall silent beside a concrete ticket hall faced with pediments and columns, and painted duck-egg blue.

Long after our train crosses back over the Arctic Circle, the last latitude at which the sun can remain in the sky for twenty-four hours non-stop, there is still no darkness. At 5 a.m. or 5 p.m., the carriages cluck-cluck through the forests and silent towns in mid-morning daylight.

Simon had not been asleep. He had been, I thought, in what psychologists call a state of hyperfocus. When Simon is thinking about something he cares for deeply—the

Cambridge–Huntingdon rail link, an unexpected digit in the Monster's Group Table, the algebra of socks —he responds as if he's flown off to stand *in* the field of the subject; he's considering the crop and its abstractions and unknown dimensions and practicalities in situ. In front of his eyes, the regular world has been replaced by a frenzy of Tuesday-afternoon arrival times, or an unexpected multiple of the number 7. A beaming grin settles on his chin.

"This *instant*," I demanded in a whisper. "What are you looking at, in your mind?"

Nothing in his face registered.

"What are you thinking? You *must* be thinking something. Is it the Monster? Your brain *can't* be empty. *Tell* me what it is."

Still, nothing.

I felt oddly frightened.

"Simon! Snap out of it! Respond!"

"Oh dear!" he sighed. In his lap was his Thomas Cook railway timetable. I noticed now that he was tapping the cover, and had probably been doing so for some time. It showed a low-angle photograph of a red train in a field of spring flowers, with glorious snow-topped mountains behind. Simon's fingertip was on the illustrative strip of route map just below the photograph, included on the cover to tell you the whereabouts of this lovely scene.

I moved his finger aside and stared at the point he'd picked out. It was a town in Austria. It was called "Rottenegg."

In the large stations, among the characters out of Gogol, Simon relaxed. What was a moment of pause for us kicked the platform into a frenzy. Women with pallets of food and plastic buckets of drink swelled out from under the station stairs and across the rails. They glanced at Simon, but he didn't look odder than their husbands. I drank hot kvass and ate a sweet chocolate pastry. Simon checked the signposts for words he could transliterate, then raced off to the toilet because he "felt ready now."

At 2 a.m., I woke as we passed through a city. Two Russians had joined our cabin, pulled down the top beds and gone to sleep. Night had come back. Simon seemed like a schoolboy again, bashful and slight, caught by the Morse code of track lamps . . .

"What are you thinking now?" I asked, feeling full of fondness.

"About socks," he whispered back.

"What about them?"

"One of the problems might be that when I take them off, I turn them inside out, then that part that has been nearest to the skin is closest to the air."

Simon's breathing became loud, almost frightened, as though he had spotted a strange animal in the room and felt trapped.

"When did people get into the habit of washing their clothes all the time?" he burst out irritably. "One reads in books people get bothered because they don't have clean underwear. One gets the impression they change it every day."

38

It is not in my nature to be concerned about trousers.

Simon

There is no conclusion to this biography—just a stepping away.

Simon's off to Boggy Bottom: the National Express 787, change at Hemel Hempstead.

("I am not prepared to put off activities!"

"No, Simon, be still. We're not finished with you yet.")

Simon's so close to a satisfying stereotype: the world-famous mathematician with electrified hair living in indescribable mess; the fallen and lonely genius, a tramp in his own home, a perfect subject for a house-clearing show in which a busybody in flyaway glasses dumps his papers in a dumpster and leaves behind a broken and weeping outcast; a tragic case. Yet every time you try to pin categories like these on Simon he steps firmly aside: he's not crazy, there's nothing tragic about him, he's definitely not poor and his life is full of purpose.

In fact, he's rushed off his feet. He's got a new newsletter to write (about a man who forces his child to eat grass because of the bus cuts); the Liberal Democrats and David Cameron to defeat; his £10,000 death-to-cars transport prize to hand out— let's hope that this time the winner won't try to superglue himself to the Prime Minister.

("Why should we hope that?"—Simon.)

Simon might have been lonely once, but politics and buses have dispensed with that: he rarely takes a journey without making a few temporary friends. Simon's lack of dejection is exhausting.

English is a bad language for gentle contentment. A glance in the thesaurus reveals smile, smirk, grin, twinkle, beam—five words to cover the full range of Simon's facial expression; and of those "smirk" is reserved for spivs about to knife you and "twinkle" employed only by novelists who use sugar tongs at tea. But look up "grief": lamentation, discontent, suffering, pain, melancholia—all the rich language that makes unhappiness so much easier a subject to write about than equanimity—*these* words go on for five pages.

Simon and I have nearly finished tidying the front room of the Excavation. I am no longer a biographer, I am a housekeeper. Next week we're hiring a weekly cleaner. The kitchen sparkles; the chest of drawers with Tango bottles glows with beeswax; the carpet emits alpine smells of Shake 'n' Vac; twenty-three six-foot piles of out-of-date timetables teeter in the corridor next to the raised Titanic Toilet, waiting for the carpenter to come on Thursday and put up wall-to-ceiling shelves in my old study.

The Excavation does not look tidy. It looks desolate.

Clear up the timetables, run a few loads of clothing through the wash, hand him over to the barber for two hours and (I'm appalled to say) Simon's character is revealed as sane, sensible and almost within grasp.

Now, could I please step aside and let him begin his day trip to Boggy Bottom?

Mathematicians who know Simon are not dismayed by his story. He may not have fulfilled his stellar potential—he's not the Newton of the twenty-first century—but he's still a genius. His performance at that recent symmetry conference in Canada was masterful. He has written numerous published and well-received papers since his supposed "collapse." It is possible the breakthrough may come any day.

The idea that Simon has altogether given up research and spends his time memorizing timetables is schadenfreude romance. Mathematicians are highly competitive, often spiteful; they enjoy spreading rumors. Simon just likes traveling on buses and trains. He has a mathematician's recall for figures, and so he can spout connection times (as far back as 1979) of the X43 Abergavenny–Brecon and the X63 Brecon–Swansea. That's not memorization, that's inevitability.

Where Simon is different from other middle-aged mathematicians is that he doesn't mope over his lost youth. He's got on with things. He's discovered what makes him content: going on day trips—

"They are *not* day trips, they are campaigning."

—going campaigning twenty-four hours a day, seven days a week—and has had the rare courage and lack of concern for public opinion to grab his delight.

"Duty!"

Duty. He doesn't want to sit all day in a neon-lit office block working out the thirteenth Fourier coefficient of a modular function on a twenty-six-dimensional hyperbolic hyperplane, thank you very much—he already knows how to do that much better than most of the rest of the people in the department anyway.

He wants fun.

"Aaah, I mean to say . . . uuugh . . . it is and it is not about fun. Oh dear. Can I go now, please?"

And fun means growing your fingernails an inch long, not cutting your hair for three years, wearing polyester trousers until the rips in the seams reach your thigh, and spending tempestuous afternoons in July dragging a bag full of railway brochures and gout pills up a hedgehog track in a Norwegian mountain forest.

"It is about fun because fun is what this Cameron government wants to destroy with our public-transport services, cutting enjoyable and vital links for the people of

Britain, leading to increased pollution, motor traffic and global warming."

Simon and I do not agree about how this biography should close any more than we agreed about how it began. Simon says I remain as I have been throughout: shallow, unreliable, obsessed with irrelevant things, obsessed with describing grime, obsessed with comical-sounding bus-stop names, a disaster for facts wherever they have the misfortune to be flushed out by me, a consistent betrayer of biographical honor.

Frédéric Chopin (another child prodigy) received a famous review from Schumann that trumpeted: "Hats off, gentlemen, a genius!"

In Simon's case this cannot any longer be said. But healthy, his mind constantly occupied, fighting an important social cause, he has no sense of loss.

("Except about government cuts to buses.")

To my mind, Simon has achieved something else which is truly important—perhaps even more so than genius.

"Hats off, gentlemen . . ."

"*Ladies* and gentlemen," he corrects, barging past.

"*Ladies* and gentlemen, hats off!" I shout after him: "There goes a happy man!"

Thank you for sending me the manuscript.
I have an incentive to deal with it quickly—
if I can do that I can throw it away.
Simon

Acknowledgments

This book could not have been written without the wizardry of Dido Davies. From helping to transcribe the sound of plastic bags and calculating the typography of a taunt (i.e., Cabbage), to rewriting practically the entire opening chapter, her energetic and inspired suggestions about structure and style, her brilliant, subtle ear for humor and pacing (and her calm encouragement in times when I've been about to give the project up in despair) have improved every page immeasurably.

Simon's brothers, Michael and Francis, his sister-in-law Amanda, and his cousins Valerie Collis and Prudence Burnett have been generous with their time and provided many of the anecdotes about Simon's family and early life (on which Simon was especially hopeless). I am also grateful to Michael for the use of the photographs of Simon and of his parents in Chapter 9, and to Amanda (in the same chapter) for the photographs of Simon's mother and of the stallion Battleaxe. Also, Vera Lucia Ottoni Guedes-Carlos Pereira-Molendini, for her stories and her ebullience: she was Simon's mother's companion, and still comes once a year to my house to cut Simon's hair.

Simon's days at Ashdown could not have been described without the enthusiasm and the superb memory of the ex-headmaster Clive Williams, and the kindness and courage of one of Simon's former (but by no means worst, Simon assures me) bullies, Malcolm Russell.

For information about Simon's days at Eton I am most indebted to William Waldegrave (now provost of the school) and to one of Simon's mathematics teachers there, Dr. Norman Routledge. Colin and Carlyn Chisholm were also helpful, though their information (including the story about Simon cutting his hair, removing too much, and trying to stick it back on with Sellotape) came too late to include in the manuscript. I hope to change that for the paperback.

Nick Wedd and Dr. Geoff Smith (who lives in a house called *Dunsummin* and is a former leader of the British Mathematics Olympiad team) provided essential information about the International Mathematics Olympiad and the childhoods of mathematical prodigies.

For their anecdotes and observations about Simon as a mathematician at Cambridge, his genius, the *Atlas* project, backgammon in the Common Room, the delightful and peculiar behavior of mathematicians in general and the Monster and Moonshine: Professor John Horton Conway, Dr. Larissa Queen, Dr. Richard Parker, Professor John McKay, Professor Koichiro Harada, Dr. Ian Grojnowski (also for his top-up lesson in Group Theory, in the tea room of the LRB Bookshop), Professor Robert Curtis, Professor Bernard Silverman, Dr. John Duncan, Professor Richard Borcherds, Professor Rob Wilson, Amanda Stagg and Dr. Ruth Williams.

Umar Salam and Dr. Graeme Mitcheson: their recommendations about how to deal with the mathematics of Group Theory, and their eagle eye for mathematical and non-mathematical errors and Cambridge misrepresentations, were invaluable.

Dr. Wajid Mannan walked across Hyde Park to my house to teach me, with patience, charm and clarity, the little I now know about how Group Theory leads to the Monster. Also, the Open University, for the excellence of its mathematics courses in Group Theory and the dedication and expertise of its tutors—I am en route to becoming its most sluggish MSc student, but this astonishing organization remains forever supportive and willing to welcome me back. While on the subject of institutions, thank you also to the NHS, because it is wonderful.

Grandmaster Ray Keene explained Simon's talent and (more interesting) his inabilities at chess. Peter Donovan and Nicholas Davidson, Simon's brilliance at bridge.

Simon has got through a vast number of tenants during his time as a landlord in Cambridge. Andrew and Anabel Turtle, Harriet Snape, Aubrey de Grey (now well known as a "gerontologist" who believes we can live until we're a thousand years old), and the late Andrew Plater and Mark Offord (Simon's current tenant, in my old rooms) have each been helpful with stories and observations.

Marianna Vintiadis reminded me of dozens of anecdotes, and analyzed them in cleverer ways than I could have managed.

Siobhan Roberts—without her energy, and her determination to get all the *Atlas* editors together for her forthcoming biography of John Conway (in which Simon will also appear, for certain), I would not have been able to write Chapter 32.

During the writing of this book, I stayed in the late Joyce Rathbone's house in Notting Hill. Though we never met, I have spent so long working beside her piano and her books that I pray some of her spirit rubbed off on me. Her goddaughter Pippa Harris was an extraordinarily generous landlady. Thank you also to Diana Berry both for her Eton contacts and for letting me use her flat in Wiltshire in the last summer of writing, when I needed a place to weep, scream and stomp; and to "Basic Dog" (Gracie) for calming my nerves by chasing anything I kicked. And to Belinda and Diana Allan, for use of their glorious house in Italy. Reading this list, I see I am a sponger.

For typing up the tapes and digital recordings of interviews, I and my hands are indebted to Clare Sproston, Ophelia Field (both of whom gave stern advice when I was sounding ridiculous or professionally incompetent) and Denise Knowelden.

Peter Straus, my agent: for his encouragement, shrewd ideas about how to present the book, and his lack of fuss. Without him there would be no contract and therefore no comfort. And Jennifer Hewson, for her constant enthusiasm and helpfulness. Nicholas Pearson, my editor at Fourth Estate in England, for his calm and patience (and excellent taste); Robert Lacey, for his unmatchable copyediting and proofreading. In America, my editors Beth Rashbaum (whose intolerance of whimsy and pages of critique improved the style in several important sections) and Ryan Doherty, and my agent George Lucas.

Thank you to my splendid and inspiring mother, Joan Brady.

And, with my love, thank you to Flora Dennis. In our endless discussions about Simon and the manuscript, Flora's fierce fight against inconsistencies, pomposities, obscurities (especially in the mathematical sections), ponderosities, purplifications, narcissisms and irrelevances has saved the story countless times. It has taken me five years to write this book, and—from Kerala to Great Snoring, Norfolk—Flora has made each one of them joyful.

Further Reading

Good Books for Group Theory and Symmetry, and Mathematical Inspiration

There are hundreds. The trick with mathematics books is to play the field. A style that one reader finds clear, evocative and memorable can be turgid, fussy or superficial for the next. Any determined Group Theory novice should have at least ten or fifteen introductory books on the subject, none of which they read thoroughly. I began my mathematics again with an old school text: Bostock, Chandler and Rourke, *Further Pure Mathematics,* which gives a straightforward, concise introduction to the basic ideas of Group Theory and the formal ways to represent them (i.e., not by squares with arms and legs, or triangles wearing bloomers). After that, rummage the bookshops and libraries. You need to be able to investigate the book close-up before you buy it. Conway has just published *The Symmetries of Things,* which looks exceptionally clear and entertaining—I'm about to start it. *Fearless Symmetry* by Avner Ash and Robert Gross is superb: casual, reflective, a balance between easy-to-read prose and mathematical exercises, it is an attempt to guide the reader toward a feel for Wiles's proof of Fermat's Last Theorem. Books like this are triumphs of intelligence and good humor. My original plan for this book was to use a similar approach, teaching myself the mathematics of the Monster and Moonshine along the way. I soon gave up. The Schaum Outline series (usually so good) is fussy, stilted and over-technical on Group Theory. I found Joe Rosen's *Symmetry Discovered: Concepts and Applications in Nature and Science* essential. It made me understand why the concept of symmetry is so important in science, when the world science is attempting to describe is, as anyone can see, blatantly asymmetrical.

Simon recommends Marshall Hall's *Theory of Groups.* He calls it "introductory." It looks terrifying. For his other mathematical interests he suggests *On Numbers and Games* (by John Conway) and *Winning Ways for Your Mathematical Plays* (Berlekamp, Conway and Guy), both classic texts that include mathematical discoveries by Simon.

Popular and recreational mathematics books are sometimes easy to understand but often not mathematically trivial, and they give Simon "new ideas to work on." He admires all those by Martin Gardner and Ian Stewart, Raymond Smullyan ("particularly his exposition of combinatory logic, *To Mock a Mockingbird*—if you don't know what combinatory logic is, don't worry—you'll find out!") and *Tracking the Automatic Ant,* by David Gale. *Prime Obsession* is another good example of popular mathematics, by the cultural journalist and novelist John Derbyshire. Written for "the intelligent and curious but non-mathematical reader," it has given Simon "a good insight into what the Riemann Hypothesis is all about." Derbyshire's *Unknown Quantity: A Real and Imaginary History of Algebra* provides an excellent introduction to the development of Simon's subject. It is just what a popular mathematics book should be: gentle, chatty, anecdotal, and full of mind-aching problems.

At the time of going to press, the editors have banned "From Sex to Quadratic Forms" as the title of Simon's paper on the mathematics of socks, which is due out later this year, in *Invitation to Mathematics.* They propose "From Gender to Quadratic Forms" instead. ("Which I think is appalling.")

Good Books About Public Transport

Simon suggests one: *Transport for Suburbia* by Paul Mees. "I can't think of any other book that has been of comparable importance to my thinking. It has launched me into a campaign to secure a Swiss-style transport system in Britain."

The books that guide Simon's broader political interests and campaign work include: *The Shock Doctrine* by Naomi Klein ("After reading it I predicted that the banking crisis would provide an opportunity for right-wing governments to do their worst"), *Cities and the Wealth of Nations* by Jane Jacobs, and Jared Diamond's *The Third Chimpanzee; Guns, Germs, and Steel;* and *Collapse.*

Simon's campaign newsletter can be found at www
.cambsbettertransport.org.uk. At the time of going to press, he's a

little behind the times. The latest edition is from last year, titled "Free Live Burials!" By cutting local bus services, Cambridge County Council is giving Cambridgeshire people "the privilege of being confined within their own homes, in some cases forever." Simon also has a regular (six-monthly) column in the *Bus Users UK Magazine* (www.bususers.org, and click on the picture of the magazine). Also, *Rail Magazine:* "I have had letters published therein."

Details of Simon's annual £10,000 Sheila McKechnie Foundation Award for improving public transport access are at www.smk.org.uk/ transport/.

Simon also contributes occasional letters (some of which can be read online) to the *Independent,* the *Guardian,* the *Camden New Journal* and *Wordways,* a magazine of recreational linguistics.